A TEXT BOOK ON
GROUPS

A TEXT BOOK ON
GROUPS
(With Solved Problem Sets)

A.N. Srivastava
Ph.D.

 **Campus Books**

A TEXT BOOK ON GROUPS

© Author

1st Edition : 2014

ISBN 81-8030-028-5

Published by :
Campus Books International
4831/24, Prahlad Street,
Ansari Road, Darya Ganj,
New Delhi - 110002
Phone : 3272541, 3257835
Fax : 91-011-3257835

Laser Typesetting by:
Stylish Laser Typesetter
Delhi

Printed at:
Roshan Offset Printers
Delhi

FOREWORD

The concept of groups can be traced back to the earlier part of the nineteenth century when it stemmed in connection with the solution of algebraic equations. At the initial stages, a group was viewed as a set of permutations with the postulate that the combination of any two of these permutations also belongs to the set. In subsequent developments, however, this definition of a group was generalized with a view to introducing the concept of an abstract group, which was defined to be a set, not necessarily of permutations, together with a method of combining its elements that is subject to a few laws or axioms.

The vast present-day knowledge and developments of the theory of groups of many different kinds are motivated, in large part, by the widely-scattered applications of these groups in numerous seemingly-unconnected areas within the disciplines of the mathematical, physical, chemical, biological and engineering sciences. The fact that groups can adequately describe such immensely important intuitive ideas as *symmetry* may be considered as one of the reasons why groups are encountered so very frequently in some of the aforementioned disciplines.

There do exist many other text-books, especially those that are published outside India, which are intended to be used by students at the under-graduate as well as graduate levels. But this book is remarkably student-friendly and it provides a systematic and well-organized introduction to the subject of groups as it is taught at the under-graduate level in all universities in India. Not only does it provide the targeted audience with the basic preliminaries about sets, relations, and functions, followed by a rather lucid and well-organized account of many different topics in the theory of groups in its seven chapters, it also includes an appendix dealing with exact sequences of groups *and* a bibliography of ten earlier works on the subjects of groups and algebra for further reading by the curious and interested student.

Judging by Dr. A.N. Srivastava's past successes and achievements, not only as a long-standing researcher and teacher of mathematics, but also as an author of other textbooks, I am happy to recommend this text-book to all undergraduate (as well as beginning graduate) students in universities in India and abroad, who find the study of groups as a requirement in their curricula.

H.M. SRIVASTAVA
(Senior) Professor of Mathematics
University of Victoria
Victoria, British Columbia V8W 3P4
Canada

PREFACE

The concept of a group is the one which pervades entire world of mathematics-pure as well as applied, and this concept arose in a natural way while considering symmetries of geometrical figures. Originally groups were confined to theory of equations, transformations and symmetries till Heinrich Weber formulated the axiomatic definition of a group in 1882 which led to the notion of an abstract group. Since then the theory of groups has grown up so much that it has found applications in coding theory, cryptography and theoretical computer science apart from providing with new techniques in exploring the various implications of symmetries in diverse situations. Owing to the simplicity of its structure, groups are now the starting point of the study of modern abstract mathematics.

The present book on groups is intended as a basic text for one semester course in modern abstract algebra at the undergraduate level; of course some of the more subtle aspects go slightly beyond that level. While there are quite a few nice books on groups, this book takes a very different approach and it has been written with two pedagogical goals in mind to offer complete undergraduate and advance graduate level topics on groups in one book and to keep the presentation student-friendly to make it the best suited book for independent study as well. The book contains much material on groups that has not previously appeared in this format.

The subject matter has been divided into seven chapters apart from Preliminaries and an appendix on exact sequence of groups. Although the book requires only a modest knowledge of set theory, number sets and matrices, for the convenience of readers, the results required from these topics have been organised in Preliminaries. The appendix introduces the readers to an advance topic—the exactness of a group sequence. Chapter 1 provides the basis on which the theory stands and it also discusses simple properties of a group based on its definition. Since the origin of groups lies in transformation, a separate chapter, Chapter 2, is devoted to permutations and groups of permutations. Group tables have been frequently used to illustrate the properties of such groups. Chapter 3 introducers the important concept of subgroups and their properties in detail. Cyclic groups play main role in developing the theory as well as applications of groups and they have been dealt completely in Chapter 4. Lattice diagrams are introduced to display complete picture of a group and all of its subgroups. Chapter 5 is devoted to cosets and Lagrange's theorem and Chapter 6 discusses normal subgroups and Quotient groups. These two chapters will be useful for those who want to go deeper into the theory. Mappings involving groups are the subject matter of Chapter 7. This chapter includes study of Isomorphism and its special forms in detail. Since one learns by doing, a large number of carefully selected solved problems follow the theory in each of the sections, and, to build up confidence review exercises have been given in the end of each chapter.

    The novel approach presented in the book is an outcome of more than two decades of my teaching abstract algebra to undergraduate students. Over these years and also during preparation of the manuscript, I have been greatly helped and benefitted by the excellent works of my senior authors. I sincerely acknowledge my indebtedness to all of them. Some of these valuable works appear in the Bibliography. My thanks are due to the publishers for their keen interest and cooperation in bringing out this book.

    I look forward receiving constructive criticism and valuable suggestions which would help improving the future editions of the book.

Pune   **A.N. Srivastava**
August 2002

CONTENTS

Foreword	(v)
Preface	(vii)
Preliminaries	(xi)

Chapter	Pages
1. Binary Operations and Groups	1
2. Permutation and Groups of Permutation	83
3. Subgroups of a Group	118
4. Cyclic Groups	151
5. Cosets and Lagrange's Theorem	184
6. Normal Subgroups and Quotient Groups	209
7. Mappings in Group Theory	228
Bibliography	262

PRELIMINARIES

The objective of this part of the book is to recall those results and various notations from the topics of sets, relations and functions which we require in our subsequent study. We have assumed that the readers are familiar with these basics of set theoretic concepts and no attempt has been made to go into the details or prove the results; although they are not prerequisits to understand the book.

1. Sets

While study of *elementary mathematics* begins with counting and measuring the objects; the study of *abstract mathematics*, such as groups, is developed with the notions of a set and parts of a set.

(*i*) *Sets and Subsets* : A *set* is a collection of distinct and disjoint (distinguishable) objects. The objects are called the *elements* or *members* of the set. The word distinct means that no two members of the set are the same and the word disjoint means that an object is either an element of the set or not an element of the set.

Sets are usually denoted by capital letters $A, X, \ldots$ and the members of a set by small letters, $a, x, \ldots$ If X is a set of which x is a member, we write $x \in X$ and we say that x *belongs to* X. If y is not a member of the set we write $y \notin X$ and we say that y *does not belong to* X.

For examples, the set of integers is denoted by $\mathbb{Z}$ (the notation $\mathbb{Z}$ comes from the word ZAHLEN, a German word meaning numbers) and the set of real numbers by $\mathbb{R}$. Thus

$$\mathbb{Z} = \{\ldots -3, -2, -1, 0, 1, 2, 3, \ldots\}$$

Instead of listing all the elements, a set can also be described by following *shorter notation*

$$\mathbb{Z} = \{n : n \text{ is an integer}\}$$

which is read by saying that '$\mathbb{Z}$ is a set of all n such that (or where) n is an integer'. Similarly

$$\mathbb{R} = \{x : x \text{ is a real number}\}.$$

Also $-2 \in \mathbb{Z}$ and $\frac{1}{3} \notin \mathbb{Z}$ while -2 and $\frac{1}{3}$ both $\in \mathbb{R}$.

Thus a set is fully described by its elements. We say that the two *sets are equal* if every element of one set is an element of the other and conversely. Thus the sets X and Y are equal, written as X = Y, whenever $x \in X \Rightarrow x \in Y$ and $y \in Y \Rightarrow y \in X$. (The *Symbol* $\Rightarrow$ is used for the word 'implies') The *symbol* $\Leftrightarrow$ which stands for 'implies and is implied by' or for 'if and only if' (in short iff) can be used to state that X = Y by writing $x \in Y \Leftrightarrow x \in Y$.

A set having only a finite number of elements is called a *finite set*, otherwise an *infinite set*. Thus $\mathbb{Z}$ and $\mathbb{R}$ are infinite sets and a set of students in a class is finite. A set having no element at all is called a *null* (or *void*) *set* and a set having only one element is called a *Singleton set*. The set of lady Presidents of India is a null set and the set of Principals in a college is a Singleton set. Null sets are denoted by the symbol ϕ (Phi) and a singleton set is denoted by mentioning the element of the set, such as {Deepanshu} or {x}.

Let X be a set then a *'part' of* X wholly lying within X, say F, is called a *subset* of X. Thus a set F is a subset of X if every element of F is an element of X and we write $F \subseteq X$ (or, conversely $X \supseteq F$) so that subset F is either fully contained in X or at the most it is equal to X.

When there is no possibility of a subset L of set X to become equal to X, we say that L is a *proper subset* of X and we write $L \subset X$, *i.e.* L is 'properly' included in X. Thus $F \subseteq X$

means $F \subset X$ or $F = X$ and $L \subset X$ means L is a subset of X such that $L \neq X$.

Any two sets A and B are equal if and only if $A \subseteq B$ and also $B \subseteq A$. Thus a set A is always a subset of its own and we have $A \subseteq A$. For example $\mathbb{Z} \subset \mathbb{R}$ (or $\mathbb{R} \supset \mathbb{Z}$, read as $\mathbb{R}$ contains $\mathbb{Z}$) and $\mathbb{Z} \subseteq \mathbb{Z}$.

(ii) *Intersection, Union and Compliment of a Set* : Intersection and union are associated with two or more sets and compliment is associated with a set and its subsets. The set of all elements which are common to two sets X and Y forms the *intersection* of X and Y. We denote this set by $X \cap Y$. Thus $X \cap Y = \{t : t \in X \text{ and } t \in Y\}$ and $u \in X \cap Y \Leftrightarrow u \in X \text{ and } u \in Y$. If nothing is common between two sets, say A and B, we write $A \cap B = \emptyset$ and say that the sets A and B are *disjoint*.

The set formed by collecting all the elements of a set X togetherwith all the elements of set B is called *union* of the sets X and B. We denote this set by $X \cup B$. Thus

$X \cup B = \{v : v \in X \text{ or } v \in B\}$ and

$\omega \in X \cup B \Leftrightarrow \omega \in X \text{ or } \omega \in B$.

If F is a subset of X then the set $X - F$ obtained by collecting all those elements of X which do not belong to F is called the *compliment* of F in X. Thus $X - F = \{x \in X : x \notin F\}$. The set $X - F$, sometime also called the *difference set*, is denoted by F' or by F^c.

(iii) *Ordered Pairs and the Cartesian Products* : We use the symbol (u, v) to denote a pair of two objects u and v. If the positions of u and v in the pair (u, v) are fixed and u, v can not be interchanged we say that (u, v) is an *ordered pair of objects* u and v. Notice that an ordered pair (u, v) is different than the set $\{u, v\}$. Two ordered pairs (u, v) and (s, t) are said to be *equal* iff $u = s$ and $v = t$.

The collection of all ordered pairs (u, v) where $u \in A$ and $v \in B$ is called *the Cartesian product* of the sets A and B. We denote this set by A × B. Thus

$$A \times B = \{(u, v) : u \in A, v \in B\}$$

If $A = \phi$ or $B = \phi$ then $A \times B = \phi$.

Examples

Let $A = \{1, 2\}$ and $B = \{a, b, c\}$, then

$$A \times B = \{(1, a), (1, b), (1, c), (2, a), (2, b), (2, c)\}$$

and $B \times A = \{(a, 1), (a, 2), (b, 1), (b, 2), (c, 1), (c, 2)\}$

so that $A \times B \neq B \times A$.

Cartesian products can also be defined when the sets A and B are equal. For example the Cartesian product of set of real numbers is the set

$$\mathbb{R} \times \mathbb{R} = \{(x, y) : x, y \in \mathbb{R}\}$$

We denote the set $\mathbb{R} \times \mathbb{R}$ by $\mathbb{R}^2$. Thus $\mathbb{R}^2$ is set of points on a plane whose co ordinates are (x, y) so that $\mathbb{R}^2$ denotes the *Cartesian plane* in two dimension. Similarly, by extending the concept, we define

$$\mathbb{R} \times \mathbb{R} \times \mathbb{R} = \{(x, y, z) : x, y, z \in \mathbb{R}\}$$

and denote it by $\mathbb{R}^3$, which is called the Cartesian plane in three dimension; and

$$\mathbb{R} \times \mathbb{R} \times \ldots \mathbb{R} \ (n \text{ times}) = \{(x_1, x_2, \ldots x_n) : x_i \in \mathbb{R}, i = 1, 2 \ldots n\}$$

denoted by $\mathbb{R}^n$ is the Carteisian plane in n-dimension or the *Euclidean n-space*.

(iv) *Set of Sets and Power Set* : A set whose elements are themselves sets is called the *set of sets*. Let A, B, C be any three sets and A denotes the collection of these sets, *i.e.*

$$\mathcal{A} = \{A, B, C\}$$

then A is a set of (three) sets. Thus 'elements' of A are themselves sets so that, for example, B ∈ A. In particular case the set of all subsets of a non-empty set is called its *power set*, denoted by letter P. Thus if X is a non-empty set then its power set, denoted by P (X), is the set of all of its subsets. Note that ϕ and $X \in P(X)$. For example if X = {a, b, c} then apart from null set ϕ and the set X itself, other subsets of X are

$$A_1 = \{a\}, A_2 = \{b\}, A_3 = \{c\},$$
$$B_1 = \{a, b\}, B_2 = \{a, c\} \text{ and } B_3 = \{b, c\}$$

so that

$$P(X) = \{\phi, X, A_1, A_2, A_3, B_1, B_2, B_3\}$$

is a set of 9 (3^2) 'elements' which themselves are sets.

2. Relations and Equivalence Relations

In Cartesian product of sets A and B, there exists *relationship between elements* of certain ordered pairs. For example if W denotes the set of all living human beings then the relationship that '*a* is the father of *b*' holds between elements of certain pairs of W × W and it does not hold true for certain other pairs. Similarly the relationships of being '*a* is greater than *b*' or '*a* is equal to *b*' are true for certain elements of pairs of $\mathbb{R} \times \mathbb{R}$.

Thus a *relation* over a set is described by a statement involving certain pairs of elements of X. A relation over a set X is denoted by the symbol ~ (or sometime by R) and thus, in terms of a set, a relation ~ over X is a subset of the Cartesian product X. When the two elements *a*, *b* of X are 'related' we write $a \sim b$. Since the relation ~ described above involves a pair of elements of set X, we call such relation a *binary relation*. Similarly we can describe a *ternary relation* involving three elements of set X. *We will restrict ourselves to binary relations.*

(i) *Special types of relations* : Let ~ be a relation over a set X.

(I) If for all $a \in X$, $a \sim a$ (*a* is 'related' to *a*) then ~ is said to be a *reflexive relation*.

(II) If for all $a, b \in X$, $a \sim b \Rightarrow b \sim a$ then ~ is said to be a *symmetric relation*.

(III) If for all $a, b, c \in X$ and $a \sim b, b \sim c \Rightarrow a \sim c$ then $\sim$ is said to be a *transitive relation*.

For example, in the set $\mathbb{R}$ the relation of equality satisfies all the above three conditions and hence it is reflexive, symmetric and transitive relation while in $\mathbb{R}$ the relation 'is less than' satisfies only the condition of transitive relation.

(*ii*) *Definition* : A relation $\sim$ over a set X is said to be an *equivalence relation* if $\sim$ satisfies all the above three conditions of reflexive, symmetric and transitive relations. Thus the relation of equality is an equivalence relation over $\mathbb{R}$ (or over any set of numbers).

(*iii*) *Partition of a set* : A set $\mathbb{P}$ of all non-empty disjoint subsets of a set X is called a *partition* of X if X is union of elements of $\mathbb{P}$.

For example, in 1 (*iv*) above the set X has partitions

$$\mathbb{P}_1 = \{A_1, A_2, A_3\}, \quad \mathbb{P}_2 = \{A_1, B_3\}$$
$$\mathbb{P}_3 = \{A_2, B_2\} \text{ and } \mathbb{P}_4 = \{A_3, B_1\} \text{ as,}$$

for example, $\quad X = A_1 \cup A_2 \cup A_3$

or $\quad X = A_1 \cup B_3$.

Similarly if $\mathbb{Z}^+$ and $\mathbb{Z}^-$ respectively denote the sets of positive and negative integers then $\mathbb{P} = \{\mathbb{Z}^+, \mathbb{Z}^-, \{0\}\}$ is a partition of the set of integers $\mathbb{Z}$.

(*iv*) *Equivalence classes* : Let $\sim$ be an equivalence relation over set X and $a \in X$. The set of those elements of X which are 'related' to a by the relation $\sim$ is called an *equivalence class of X determined by a*. We denote this set by $\bar{a}$ or by E_a. Thus $\bar{a} = \{x \in X : x \sim a\}$ and $\bar{a}$ is a subset of X and $a \in \bar{a}$, as $\sim$ is reflexive.

Since equality relation in $\mathbb{R}$ is an equivalence relation, the equivalence class determined by its element, say 2, is given by

$$\bar{2} = \{x \in \mathbb{R} : x \sim 2\}$$
$$= \{2\}, \text{ as only 2 is equal to 2.}$$

We state following *fundamental theorems* on equivalence relations.

Theorem 1 : If ~ is an equivalence relation over set X and $a, b \in X$ then

(i) equivalence classes $\bar{a}, \bar{b}$ are non-empty.

(ii) If $a \sim b$ then $\bar{a} = \bar{b}$.

(iii) If $\bar{a} \neq \bar{b}$ then $\bar{a} \cap \bar{b} = \phi$

Theorem 2 : The set of equivalence classes determined by all the elements of a set X is a partition of X. Thus $\mathbb{P} = \{\bar{u} : u \in X\}$ of all equivalence classes with regard to an equivalence relation ~ over X is a partition of X.

In other words, any equivalence relation ~ over a set X partitions X into (mutually disjoint) equivalence classes. Let $X = \{a, b, c\}$ and $\bar{a}$ be the equivalence class determined by a with regard to an equivalence relation over X; and similarly $\bar{b}, \bar{c}$ then we have the following diagram for X :

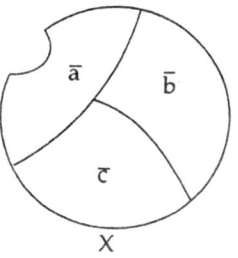

Conversely : Any partition of X into non-empty subsets, induces an equivalence relation in X, for which these subsets are the equivalence classes.

The set of equivalence classes obtained from an equivalence relation over a set X is said to be the *Quotient set* of X. We denote this set by $\overline{X}$. Thus if $X = \{a, b, c\}$ then $\overline{X} = \{\bar{a}, \bar{b}, \bar{c}\}$.

(v) *The set of integers* : In our subsequent discussions, the set which appears frequently is the set of integers $\mathbb{Z} = \{\ldots -3, -2, -1, 0, 1, 2, 3, \ldots\}$. We recall some of its properties :

(I) An integer i is said to be *divisible* by a non-zero integer j if $i = jk$ where k is some integer and we say i to be a *multiple* of j and j to be a *divisor* of i; denoted by j/a (read as j divides i).

(II) *Division algorithm for* $\mathbb{Z}$: Let a and b be any integers such that $b \neq 0$. Then there exists unique integers q and r such that $a = bq + r$, where $0 \leq r < |b|$. r is called the *remainder*.

In particular if $b > 0$ then, from above, $a = bq + r$ where $0 \leq r < b$.

A positive integer $p \neq 1$ is said to be a *prime number* if it has no positive integral factors except 1 and itself. Examples are 2, 3, 5, 7, ...

The *highest common factor* (HCF) or the *greatest common divisor* (GCD) of two positive integers l and m is the largest positive integer h which is a factor of both, l and m. Thus $h =$ GCD $\{l, m\}$ if h/l and h/m and if k is a common divisor of l and m then k/h. If either l or m is zero, their GCD is the non-zero integer. If GCD $\{l, m\} = 1$, then l and m are said to be *co prime numbers*.

The *least common multiple* (LCM) of two positive integers l and m is the smallest positive integer c which is a multiple of both, l and m. If l and m are co prime then their LCM, $c = lm$.

(III) *Congruence of integers* : The concept of congruence, developed by Gauss (1777-1855), plays an important role in the study of integers and is given as follows :

Let m be a positive integer and x, y be any two integers then we say that *x is congruent to y modulo m* (or, in short, mod m) if m is a divisor of $x-y$ and we write it as $x \equiv y \pmod{m}$. Thus $a \equiv b \pmod{m}$ implies that $a, b \in \mathbb{Z}$, $m \in \mathbb{Z}^+$ and $a - b = km$, where k is same integer.

For example $24 \equiv 4 \pmod 5$, $-6 \equiv 2 \pmod 4$ etc.

For a fixed positive integer m, the relation of 'Congruence modulo m' is an equivalence relation over the set of integers $\mathbb{Z}$ and this equivalence relation divides $\mathbb{Z}$ in m distinct equivalence classes. The equivalence classes are called the *residue classes* (or the congruence classes) modulo m. For example if $x \in \mathbb{Z}$ then the *residue class of x* mod m, denoted by $\bar{x}$, is given by

$$\bar{x} = \{x + km : k \in \mathbb{Z}\}$$
$$= \{\ldots -3m + x, -2m + x, -m + x, x, x + m, x + 2m, x + 3m, \ldots\}$$

If $x = 2$ and $m = 4$ then residue class of 2 mod 4 is given by

$$\bar{2} = \{2 + 4k : k \in \mathbb{Z}\} = \{\ldots -6, -2, 2, 6, 10, \ldots\}$$

(IV) *Algebra of congruence :* We can add, subtract or multiply any two congruences mod m and hence sum and product of two residue classes mod m can be defined as follows : Let $\bar{a}$ and $\bar{b}$ be residue classes of integers a and b mod m respectively then

$$\bar{a} + \bar{b} = \overline{a+b} = \bar{r} \qquad \ldots(1)$$

where r is the least non-negative remainder when $a + b$ is divided by m. Since $a, b \in \mathbb{Z}$ and $a + b = b + a$,

By (1) $\bar{a} + \bar{b} = \overline{b+a} = \bar{b} + \bar{a}$, which implies that the sum of residue classes obey commutative law. Similarly if $\bar{a}, \bar{b}, \bar{c}$ denote the residue classes of integers a, b, c mod m respectively then their sum satisfy associative law, *i.e.*

$$\bar{a} + (\bar{b} + \bar{c}) = (\bar{a} + \bar{b}) + \bar{c}.$$

The product of residue classes $\bar{a}$ and $\bar{b}$ is defined by

$$\bar{a}.\bar{b} = \overline{ab} = \bar{r} \qquad \ldots(2)$$

where r is the least non-negative remainder when product ab is divided by m. Since $a, b \in \mathbb{Z}$ and $a \cdot b = b \cdot a$, by (2)

$$\bar{a}.\bar{b} = \overline{ba} = \bar{b}.\bar{a},$$

so that the product of two residue classes obey commutative law. Similarly the product of residue classes obey associative law as well, *i.e.*

$$\bar{a}.(\bar{b}.\bar{c}) = (\bar{a}.\bar{b}).\bar{c}$$

(V) *Addition and multiplication modulo a positive integer m :* Exactly like addition and multiplication of any two residue classes of integers mod m, we define a *new type of addition and multiplication* for any two integers a and b *as follows* :

The *addition modulo* m, denoted by $+_m$, for any two integers a and b is defined as, $a +_m b = r$, where r is the least non-negative remainder when the ordinary sum of a and b, $a + b$, is divided by m. Examples are

$$6 +_2 9 = 1, \quad 5 +_4 7 = 0, \quad -14 +_4 4 = 2, \text{ etc.}$$

From definition of $+_m$, it follows immediately that :

(i) $a +_m b = b +_m a$ (Commutativity of $+_m$)

(ii) $a +_m (b +_m c) = (a +_m b) +_m c$ (Associativity of $+_m$), where $a, b, c \in \mathbb{Z}$, and

(iii) $a +_m b \equiv a + b \pmod{m}$

The *multiplication modulo* m, denoted by $\times_m$, for any two integers a and b is defined as $a \times_m b = r$, where r is the least non-negative remainder when the ordinary product of a and b, ab, is divided by m. Examples are

$$6 \times_2 9 = 0, \quad 5 \times_4 7 = 3, \quad \text{etc.}$$

From definition of $\times_m$, it follows that :

(iv) $a \times_m b = b \times_m a$ (Commutativity of $\times_m$)

(v) $a \times_m (b \times_m c) = (a \times_m b) \times_m c$; $a, b, c, \in \mathbb{Z}$ (Associativity of $\times_m$) and

(vi) $a \times_m b \equiv ab \pmod{m}$.

Congruence, residue classes, addition modulo m and multiplication modulo m are parts of *'modular arithmetic'*.

3. Functions or Mappings

(i) *Definitions* : Let X and Y be any sets and f be a rule which assigns to each element f X a unique element of Y, then the rule of is called a *function* (or a *mapping*) from X to Y. We denoted it by

$$f : X \to Y.$$

If x is an element of X then the map f will assign an element of Y to x. This element of Y is denoted either by xf or by $f(x)$. *In this book we will use the notation xf except in the case when product (composition) of two maps are considered;* in that case we will adopt the notation $f(x)$. The element xf is called the *image of x* under f. The set X is called the *domain* of f and the set of all images xf, for all $x \in A$, is called the *range* (or *range set*) of f. We will denote the range of f by Xf (Instead of $f(X)$, commonly used in calculus) and read Xf as '*X under f*'. The set Y is called the *co-domain of f*. Thus $Xf = \{xf : x \in X\}$ and Xf is a subset of the co-domain Y.

In terms of relation, we define a function f as follows :

A map from X to Y is a relation f satisfying :

(a) domain $f = X$,

(b) for each $x \in X$, there exists a unique $y \in Y$ such that the ordered pair $(x, y) \in f$.

(ii) *Equality of mappings and types of mappings :* Two mapping f and g are *equal*, shown as $f = g$, if following three conditions are satisfied :

(a) f and g have the same domain

(b) f and g have the same co-domain, and

(c) for all x belonging to domain

$xf = xg$

Notice that the definition of a function does not require that distinct elements x_1 and x_2 of domain have different images in the co-domain; instead it only requires that each element x of domain has a unique image in the co domain. This leads to the following *special types* of mappings :

(I) The map $f : X \to Y$ is said to be *one-to-one* (or *injective*) if whenever x_1 and x_2 are distinct elements of X, $x_1 f$ and $x_2 f$ are distinct elements of Y, *i.e.* f is one-to-one if, $x_1 f = x_2 f \Rightarrow x_1 = x_2$ for all $x_1, x_2 \in X$. We will use this *condition* to show if a map is one-to-one. If the condition is not satisfied by a map f, we say that f is *many-one* mapping.

(II) The map $f : X \to Y$ is said to be *onto* (or *surjective*) if range set Xf is equal to the co-domain Y. Thus f is onto then

$Xf = Y$, and then for each $y \in Y$, there exists $x \in X$ such that $y = xf$.

If $Xf \subset Y$ then the map f is said to be *into* (Note that $Xf \not\supset Y$).

(III) The map $f : X \to Y$ is said to be a *one-to-one correspondence* (or *bijective*) if it is one-to-one and onto, and then the sets X and Y are said to be in one-to-one correspondence. One-to-one onto maps are significantly important in the sense that they have very useful properties, one of which is that inverse maps for them can be defined.

(iii) *Inverse mappings* : If $f : X \to Y$ is one-to-one onto map then its inverse, denoted by f^{-1}, is the map $f^{-1} : Y \to X$ such that

$$yf^{-1} = x \Leftrightarrow y = xf$$

It can be easily shown that the map f^{-1} is also one-to-one onto and hence its inverse can be defined, which is f itself.

(iv) *Direct and inverse images* : If $f : X \to Y$ is a map, A and B are subsets of X and Y respectively, then $(A)f$ is called the *direct image of* A under f, given by

$$(A)f = \{y \in Y : \text{there exists } x \in A \text{ such that } y = xf\}$$

and hence $(A)f$ is the set of all images of the elements of $A \subset X$. Similarly the set $(B)f^{-1}$ is called the *inverse image of* $B \subset Y$ under f and is given by

$$(B)f^{-1} = \{x \in X : xf \in B\}$$

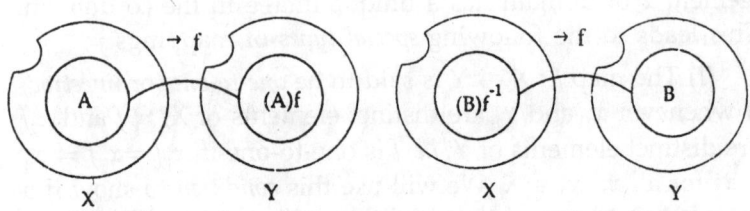

direct image   inverse image

Notice the difference between inverse of a map $f : X \to Y$ and the inverse image of a subset of Y under f.

(*v*) *Identity mappings* : Let X be a set and $I : X \to X$ be a map from X to X. Then I is said to be the *identity mapping* on X if for all $x \in X$, $xI = x$.

By its definition, I is one-to-one onto and its inverse, $I^{-1} = I$.

We will denote the set of all mappings from X to X by F_x and the set of all one-to-one onto mappings from X to X by B_x. Thus $I \in F_x$ as well as $I \in B_x$.

(*vi*) *Product of mappings* : Let $f : X \to Y$ and $g : Y \to Z$ be mappings, then the mapping $fg : X \to Z$ defined by $x\,(fg) = (xf)\,g$ for all $x \in X$ is called the *product* of two mappings f and g. Product of maps is also called *composition of maps* and instead of denoting it by fg, it is also denoted by gof, o being the composition (product). In the definition, $x\,(fg) = (xf)\,g$ notice that $xf \in Y$ and hence $(xf)\,g \in Z$. In case of product of mappings it is convenient to follow the notation $f(x)$ for xf and hence we define the product map fg as :

$$(gof)\,(x) = g\,(f\,(x))$$

where $x \in X$, $f(x) \in Y$ and $g\,(f(x)) \in Z$.

When it comes to product of mappings, we will follow this *second notation* throughout this book.

Note that if *gof* is defined then *fog* need not be defined.

Associativity and commutativity of product of mappings : By similarly defining product of three maps $f : X \to Y$, $g : Y \to Z$ and $h : Z \to W$ as the mapping $(fgh) : X \to W$.

(*i.e.* $ho\,(gof) : X \to W$), we can prove that $(fg)\,h = f\,(gh)$ (*i.e.* $ho\,(gof) = (hog)\,of$), which is associative law for the product of mappings. The product of mappings, in general, does not obey commutative law; *i.e.* $gof \neq fog$ in general.

(*vii*) *Product of one-to-one onto mappings* : We have following results, which can be easily verified by taking examples, for the product of maps $f : X \to Y$, $g : Y \to Z$ and $h : Z \to W$ where all three maps f, g and h are one-to-one onto.

(I) If f and g are both one-to-one then fg is one-to-one.

(II) If f and g are both onto then fg is also onto.

Combining above two, we get

(III) If f and g are both one-to-one onto maps then their product fg is also one-to-one onto maps.

(IV) $$ff^{-1} = f^{-1}of = I_X \text{ and}$$
$$f^{-1}f = f^{-1}of = I_Y,$$

where I_X is identity map on X and I_Y is identity map on Y.

(V) Product of maps f, g, h satisfy associative law, $f(gh) = (fg)h$. Proceed by considering $(ho\,(gof))(x) = h\,((gof)\,x)$ and apply 3 (vi).

(VI) In general, product of f and g does not satisfy commutative law, i.e.

$$fg \neq gf \quad (i.e.\ gof \neq fog)$$

We will give a particular example to illustrate this :

Let $X = \{a, b, c, d\}$ and f_1 and f_2 are one-to-one onto maps both from X to X itself, defined as follows :

$f_1(a) = b, f_1(b) = c, f_1(c) = a$ and $f_1(d) = d$, and $f_2(a) = a$, $f_2(b) = c, f_2(c) = b, f_2(d) = d$.

Then, by definition of product

$$(f_1 f_2)(a) = f_1(f_2(a)) = f_1(a) = b,$$
and $$(f_2 f_1)(a) = f_2(f_1(a)) = f_2(b) = c,$$

showing that $f_1 f_2 \neq f_2 f_1$.

$(viii)$ Two sets X and Y which are in one-to-one correspondence are said to have the same *cardinal numbers* and we write card X = card Y. Thus for having same *cardinality*, there must exist a function $f : X \to Y$ between the sets X and Y which is both, one-to-one and onto. The two sets are then called *similar sets* or *equipotent sets*. Property of being similar is a relation between sets which satisfies all conditions of an equivalence relation. If a set X is in one-to-one correspondence with the set $\{1, 2, \ldots n\}$ of first n natural numbers then X is said to be a *finite set* and if X is in one-to-one correspondence with the set $\mathbb{N}$, set of all natural numbers, then X is said to be *countably infinite set*. Thus the concept of cardinality play an important role in comparing different sets.

1 BINARY OPERATIONS AND GROUPS

1.1 Binary Operation

A prerequisite to define and study groups is to understand a binary operation (in short b.o.) which is a special type of function defined as follows :

Definition 1.1.1 : Let A be a non-empty set. A binary operation (*b o*) or a binary function over A is a function f from the set $A \times A$ to A. Symbolically, f is a *b.o.* if f is a function, $f : A \times A \to A$, where $A \times A$ is Cartesian product of set A. (See prelim. 1 (*iii*)

(*i*) According to definition of a function, a b.o. will carry each ordered pair (a, b) belonging to the set $A \times A$ to a unique element $(a, b)f$ belonging to the set A. (We will use the notation xf in place of $f(x)$. xf is read as 'x under f').

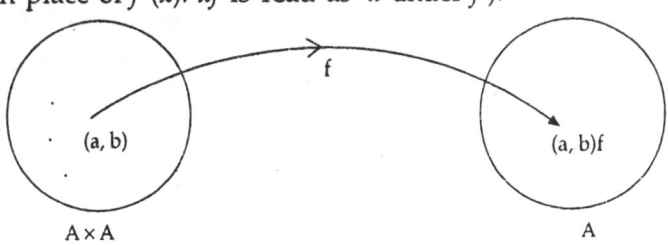

$A \times A$ A

(Set of all ordered pairs of elements of A).

Depending on the rule f, $(a, b)f$ will have a unique value belonging to A.

(*ii*) The domain of b.o. f is $A \times A$ and its co-domain is A.

(iii) In general a b.o. will be denoted by ∗ in place of f (or g, h...) to differentiate it from 'ordinary functions'. Thus ∗ will mean a b.o. over A and ∗ : A × A → A, and we will write that ∀ (a, b) ∈ A × A, there exists a unique element (a, b) ∗ ∈ A.

(iv) The value of (a, b) ∗ ∈ A will be denoted by $a * b$. Thus, if ∗ = + (addition), then the element (a, b) is carried over to $(a, b) * = (a, b)+ = a + b$.

Similarly if ∗ = −, × or ÷ (subtraction, multiplication or division) then (a, b) ∗ will mean

$$(a, b) - = a - b$$
$$(a, b) \times = a \times b$$

and $\qquad (a, b) \div = a \div b$ respectively.

(v) Like other functions a b.o. may also be one to one, many one, onto or into.

Note : (i) The definition of a b.o. can be widened by introducing the terms *internal b.o.* and *external b.o.* The map f : A × A → A is internal b.o. over A and the map f : A × B → B, where B is a non-empty set, is an external b.o. on B (over set A). The notion of external b.o. appears at more advanced stages of study. For our purpose, a b.o. will mean an internal b.o.

(ii) Whenever a function f is a b.o. over set A, we also say that f is a *clsoed operation* or that the set A is closed with respect to operation f. Thus a b.o. f over a set is also characterised by saying that the operation f is a closed operation.

1.1.1 Illustrative Examples

1. (i) In the definition 1.1.1, take f as +, set A = $\mathbb{Z}^+$ then the rule of addition + over the set $\mathbb{Z}^+$ is a b.o. because + : $\mathbb{Z}^+$ × $\mathbb{Z}^+$ → $\mathbb{Z}^+$. The reason being that for any m, n ∈ $\mathbb{Z}^+$, the ordered pair (m, n) ∈ $\mathbb{Z}^+$ × $\mathbb{Z}^+$ will be carried over by + to $(m, n) + = m + n$, which is a unique element of $\mathbb{Z}^+$. In particular, the ordered pair $(3, 5)$ → $(3, 5) + = 3 + 5 = 8$. Notice that the b.o. + over $\mathbb{Z}^+$ is many one into as well.

Binary Operations and Groups

Similarly the rule of addition + is b.o. over the sets $\mathbb{Z}, \mathbb{Q}, \mathbb{R}$ and $\mathbb{C}$. + is not a b.o. over $\mathbb{Z} - \{0\}$, why?

(ii) The rule of subtraction—is not a b.o. over $\mathbb{Z}^+$ as $(3, 5) \in \mathbb{Z}^+ \times \mathbb{Z}^+$ is carried over by - to the element $(3, 5) - = 3 - 5 = -2$, which does not belong to $\mathbb{Z}^+$. Thus $- : \mathbb{Z}^+ \times \mathbb{Z}^+ \not\to \mathbb{Z}^+$. But $- : \mathbb{Z} \times \mathbb{Z} \to \mathbb{Z}$, i.e. subtraction is a b.o. over $\mathbb{Z}$. Similarly $- : \mathbb{Q} \times \mathbb{Q} \to \mathbb{Q}$, $- : \mathbb{R} \times \mathbb{R} \to \mathbb{R}$ and $- : \mathbb{C} \times \mathbb{C} \to \mathbb{C}$.

(iii) The rule of multiplication $\times$ carries over the element $(m, n) \in \mathbb{Z}^+ \times \mathbb{Z}^+$ to a unique element $m \times n \in \mathbb{Z}^+$, thus $\times : \mathbb{Z}^+ \times \mathbb{Z}^+ \to \mathbb{Z}^+$, i.e. $\times$ is a b.o. over $\mathbb{Z}^+$.

Similarly $\times : \mathbb{Z} \times \mathbb{Z} \to \mathbb{Z}$, $\times : \mathbb{Q} \times \mathbb{Q} \to \mathbb{Q}$, $\times : \mathbb{R} \times \mathbb{R} \to \mathbb{R}$ and $\times : \mathbb{C} \times \mathbb{C} \to \mathbb{C}$.

(iv) The rule of division $\div$ is neither a b.o. over the set $\mathbb{Z}^+$ nor over the set $\mathbb{Z}$, i.e. $\div : \mathbb{Z}^+ \times \mathbb{Z}^+ \not\to \mathbb{Z}^+$ and $\div : \mathbb{Z} \times \mathbb{Z} \not\to \mathbb{Z}$ as, for example, in the first case the pair $(2, 3) \in \mathbb{Z}^+ \times \mathbb{Z}^+$ is carried over by $\div$ to the element $(2, 3) \div = 2 \div 3$, which does not belong to $\mathbb{Z}^+$.

Even $\div : \mathbb{R} \times \mathbb{R} \not\to \mathbb{R}$, for the pair $(1, 0) \in \mathbb{R} \times \mathbb{R}$ is carried over by $\div$ to $(1, 0) \div = 1 \div 0 = \infty$, which does not belong to $\mathbb{R}$. Of course,

$$\div : \mathbb{R}^* \times \mathbb{R}^* \to \mathbb{R}^*, \quad \div : \mathbb{R}^+ \times \mathbb{R}^+ \to \mathbb{R}^+,$$

$$\div : \mathbb{C}^* \times \mathbb{C}^* \to \mathbb{C}, \quad \text{Note that } \div : \mathbb{R}^- \times \overline{\mathbb{R}} \not\to \overline{\mathbb{R}}.$$

2. Generalized Operations

New operations can be defined by combining two or more of the four arithmetical operations. For example, if for $m, n \in \mathbb{Z}$ we define $m * n$ as, $m * n = m + n - mn$, then the rule $*$ combines $+, -$ and $\times$. Such operations are called generalized operations and they occur frequently while studying binary operations or groups. Let us look at some generalized operations:

(i) The above generalized rule $*$ defined by $m * n = m + n - mn$ is a b.o. over the set $\mathbb{Z}$, because for any two elements

$m, n \in \mathbb{Z}$, $m + n - mn$ also belongs to $\mathbb{Z}$, i.e. $* : \mathbb{Z} \times \mathbb{Z} \to \mathbb{Z}$ but $* : \mathbb{Z}^+ \times \mathbb{Z}^+ \nrightarrow \mathbb{Z}^+$ as the ordered pair $(2, 2) \in \mathbb{Z}^+ \times \mathbb{Z}^+$ is carried over by $*$ to the element $2 + 2 - 2 \times 2 = 0$, which does not belong to $\mathbb{Z}^+$.

(ii) The generalized operations defined by $a * b = \frac{1}{2}(a + b + ab)$, $a *' b = \frac{1}{3}(a + b)$, $a *'' b = |a| b$ are binary operations over the sets $\mathbb{Q}$, $\mathbb{R}$.

We now consider some sets which have their elements other than numbers :

3. Over the set of all vectors $\vec{V}$, the scalar (or dot) product is not a b.o. but the vector (or cross) product is a b.o., as for any two vectors $\vec{a}, \vec{b} \in \vec{V}$, $\vec{a}.\vec{b} \notin \vec{V}$ and $\vec{a} \times \vec{b} \in \vec{V}$.

4. Over the power set P (A) of a non-empty set A, the operations of union ($\cup$) and intersection ($\cap$) are binary operations. The difference $X - Y$ is yet another b.o. over P (A), where $X, Y \in P(A)$.

5. Consider the set F_A of all mappings of A into A. Over F_A, we define the composition of two mappings f and g by

$$(f * g) x = f(g(x))$$

Then $*$ is a b.o. over F_A, (we will use the notation $f(x)$, for xf, when the set F_A is under consideration). Since $f : A \to A$ and $g : A \to A$, by definition of composition, as defined above, $(f * g) : A \to A$ so that $f * g \in F_A$.

6. Let $\mu_{(m \times n)}(\mathbb{R})$ and $\mu_{(m \times n)}(\mathbb{C})$ be the sets of all $m \times n$ matrices with their elements as real numbers and complex numbers respectively. Let $*$ denote sum of two such matrices, then $*$ is a b.o. over $\mu_{(m \times n)}(\mathbb{R})$ and $\mu_{(m \times n)}(\mathbb{C})$. similarly if $*'$ denote product of two such matrices from $\mu_{(m \times n)}(\mathbb{R})$ and $\mu_{(m \times n)}(\mathbb{C})$ respectively then $*'$ is also a b.o. over them.

1.1.2 Commutative and Associative Binary Operations

Definition 1.1.2 : A b.o. over a set A is said to be

Binary Operations and Groups

commutative if $a * b = b * a$, true for all the elements $a, b \in A$.

Definition 1.1.3 : A b.o. over a set A is said to be associative if $a * (b * c) = (a * b) * c$, true for all the elements $a, b, c \in A$.

Note : (i) By repeated application of associative law, it is obvious that different ways of putting brackets in the composite element $a_1 * a_2 * \ldots * a_n$ give rise to the same element as long as we preserve the sequence in which the elements appear. Thus

$$a_1 * a_2 * \ldots * a_n = a_1 * (a_2 * a_3 * \ldots * a_n)$$
$$= (a_1 * a_2 * \ldots * a_{n-1}) * a_n = \ldots$$

This is *the generalized associative law*.

(ii) If a b.o. * over set A is both associative and commutative then the order in which the elements in the 'product' $a_1 * a_2 * \ldots * a_n$ appear is immaterial.

Definition 1.1.4 : The symbol a^n : When n is a positive integer, we write $a * a * \ldots * a$ (n times) as a^n. Thus $a^3 = a * a * a$. Let $* = +$, then $a^3 = a + a + a = 3a$ and $a^n = a + a + \ldots + a$ (n times) $= na$. Similarly if $* = \cdot$ then $a^3 = a \cdot a \cdot a$ will carry the usual meaning of a^3.

We will give wider meaning to the symbol a^n when n is any integer in chapter 4.

1.2 Binary Operation Tables

A b.o. * over a small finite set can also be displayed in a tabular form. The elements of the set are written along the top-most row and also along the extreme left corner of the table; the intersection of the corresponding row and column of the elements give us the value of the operation.

Let $S = \{a, b, c\}$ be a set having three elements. Then one of the binary operations over S can be put in the following table :

B.O.	a	b	c
a	a	c	c
b	a	c	a
c	b	c	c

Table 1.2.1

In general, a b.o. table is read as follows :

(ith entry along extreme left corner) * (jth entry along topmost row) = entry in the ith row and jth column of the table.

Thus, $b * c = a$, $c * b = c$ etc., and $(c * a) * (b * b)$.
$$= b * c$$
$$= a$$

Binary operation tables are also called *multiplication or composition tables or Cayley's tables*.

1.2.1 Total Number of Binary Operations

It is possible to formulate the total number of binary operations which are possible over a finite set S. In a b.o. table for set with 3 elements S = {a, b, c}, total 9 places are available to be filled from amongst the elements of S, and each place can be filled in 3 different ways by picking up any of the 3 elements a, b or c. Since each way of filling a place gives rise to a different b.o., thus :

Total number of binary operations possible over a set of 3 elements.

= Total number of different ways of filling the 9 places

= 3.3.3 ... 9 times

= $3^9 = 3^{3^2}$

B.O. *	a	b	c
a	·	·	·
b	·	·	·
c	·	·	·

Principal diagonal

Table 1.2.2

In general, if a finite set S has n elements then the total number of binary operations possible over it will be given by the *formula* n^{n^2}.

1.2.2 Commutative Binary Operation Tables

If a b.o. defined over a set is commutative as well, then since $a * b = b * a$, the b.o. table will be symmetrical about the principal diagonal. In that case, from table 1.2.2, total number of places to be filled by the elements of the set $S = \{a, b, c\}$ will be equal to the number of places above (or below) the principal diagonal + places lying along it, which are

$$\frac{1}{2}(3^2 - 3) + 3 = \frac{3(3+1)}{2}$$

Then, as in the previous case, total number of commutative binary operations over S

$$= 3.3.3. \ldots \frac{3(3+1)}{2} \text{ times}$$

$$= 3^{\frac{3(3+1)}{2}}$$

Thus, in general, if a finite set S has n elements then the total number of binary operations satisfying commutative law over S will be $n^{\frac{n(n+1)}{2}}$.

1.2.3 Illustrative Examples

1. Over a set of 4 elements, 4^{4^2} i.e. 4^{16} binary operations can be defined. Out of 4^{16} possible binary operations, $4^{\frac{4(4+1)}{2}}$ i.e. 4^{10} binary operations will satisfy the commutative law. In example 9 below, table 1.2.3 is a commutative b.o. table and table 1.2.4 is a non-commutative b.o. table.

2. The b.o. + (addition) over the set $\mathbb{Z}$ is both cummutative and associative because for any $l, m, n \in \mathbb{Z}$, we have $l + m = m + l$ and $l + (m + n) = (l + m) + n$, while the b.o. − (subtraction) over $\mathbb{Z}$ is neither commutative nor associative, for $l - m \neq m - l$ and $l - (m - n) \neq (l - m) - n$. The b.o. ÷ (division) over $\mathbb{R}^*$ is also neither commutative nor associative.

3. The generalized operation $a * b = a + b - ab$, which is a b.o. over $\mathbb{Z}$ is both commutative as well associative, while

the generalized operation $a *' b = ab + 1$ which is a b.o. over $\mathbb{Q}$ is commutative but not associative.

For any $l, m, n \in \mathbb{Z}$, $l * m = l + m - lm$
$$= m + l - ml = m * l, \text{ and}$$
$$l * (m * n) = l * (m + n - mn)$$
$$= l * \alpha \quad (\alpha = m + n - mn)$$
$$= l + \alpha - l\alpha$$
$$= l + m + n - mn - lm - ln + lmn$$

Similarly $(l * m) * n$
$$= (l + m - lm) * n$$
$$= \beta * n \quad (\beta = l + m - lm)$$
$$= \beta + n - \beta n$$
$$= l + m - lm + n - ln - mn + lmn$$
$$= l * (m * n)$$

For any $p, q, r \in \mathbb{Q}$,
$$p *' q = pq + 1 = qp + 1 = q *' p$$
and $\quad p *' (q *' r) = p *' (qr + 1)$
$$= p(qr + 1) + 1 = pqr + p + 1$$
while $\quad (p *' q) *' r = (pq + 1) *' r$
$$= (pq + 1) r + 1$$
$$= pqr + r + 1$$
showing that $p *' (q *' r) \neq (p *' q) *' r$

4. (i) The b.o. $*$ over $\mathbb{R} - \{0\}$ defined by $x * y = |x| y$ satisfies associative law but not commutative law.

(ii) The b.o. $*$ over $\mathbb{R} - \{0\}$ defined by $x * y = |x - y|$ satisfies commutative law and not the associative law.

(check it yourself)

5. Let $\mu_{n \times n}(\mathbb{R})$ be the set of all $n \times n$ square matrices with elements as real numbers and let $*$ defines the product of any two such matrices then since for any $A, B, C \in \mu_{n \times n}(\mathbb{R})$,

Binary Operations and Groups

$$A * B \neq B * A \quad \text{and} \quad A * (B * C) = (A * B) * C$$

The product of matrices over $\mu_{n \times n}(\mathbb{R})$ satisfies associative law but not the commutative law. What you have to say for the operations of addition and subtraction over $\mu_{n \times n}(\mathbb{R})$?

6. Consider the power set P (A) of a non-empty set A and the binary operations of $\cup$ (union) and $\cap$ (intersection) over it. From knowledge of sets, we know that for any X, Y, Z $\in$ P (A),

$$X \cup Y = Y \cup X \quad \text{and} \quad X \cup (Y \cup Z) = (X \cup Y) \cup Z;$$
$$X \cap Y = Y \cap X \quad \text{and} \quad X \cap (Y \cap Z) = (X \cap Y) \cap Z$$

Thus the binary operations of union and intersection over the set P (A) satisfy both, commutative and associative law.

7. The vector product of any two vectors which is a b.o. over the set of all vectors $\vec{V}$ does not satisfy, in general, the commutative as well as associative law. For any $\vec{a}, \vec{b}, \vec{c} \in \vec{V}$, we know that $\vec{a} \times \vec{b} \neq \vec{b} \times \vec{a}$ and $\vec{a} \times (\vec{b} \times \vec{c}) \neq (\vec{a} \times \vec{b}) \times \vec{c}$.

8. Over the set F_A of all mappings of A into A, the composition of two mappings $*$ is a b.o. over F_A which satisfies the associative law but, in general, not the commutative law, i.e. for any $f, g, h, \in F_A$ (See, Prelim. 3 (vii), $f * (g * h) = (f * g) * h$, and

$$f * g \neq g * f$$

9. Consider, the following two binary operation tables defined over a set S = {a, b, c, d} of 4 elements :

*	a	b	c	d
a	a	b	c	d
b	b	d	a	c
c	c	a	d	b
d	d	c	b	a

Table 1.2.3

*	a	b	c	d
a	a	b	c	d
b	b	a	c	d
c	c	d	c	d
d	d	c	c	d

Table 1.2.4

From table 1.2.3,

$$c * (b * d) = c * c = d$$
and
$$(c * b) * d = a * d = d,$$
i.e.
$$c * (b * d) = c * (b * d).$$

From table 1.2.4,

$$c * (b * d) = c * d = d,$$
and
$$(c * b) * d = d * d = d,$$
i.e.
$$c * (b * d) = (c * b) * d.$$

Similarly by taking any 3 elements, we can show that the b.o. * defined in table 1.2.3 satisfies both commutative as well as associative law and the b.o. * defined in table 1.2.4 satisfies associative law but not the commutative law.

1.3 Identity and Inverse Elements

We define and illustrate two important concepts of identity and inverse elements wrt an operation.

Definition 1.3.1 : Let A be a non-empty set and * be a b.o. defined over A. An element e_L is said to be *left identity* wrt * if $e_L * a = a$, for all $a \in A$; and an element e_R is said to be *right identity* wrt * if $a * e_R = a$, for all $a \in A$.

Notice that e_L appears in the left while e_R appears in the right of their defining equations.

Definition 1.3.2 : Whenever $e_L = e_R$, the common value is denoted by e and is called *identity element* wrt *.

Thus the element e is identity element wrt * defined over A if $e * a = a * e = a$, true for all elements $a \in A$.

Definition 1.3.3 : An element a'_L (read as a dash L) is said to be *left inverse element* of an element $a \in A$ wrt * if $a'_L * a = e_L$; and element a'_R (read as a dash R) is said to be *right inverse element* of $a \in A$ if $a * a'_R = e_R$.

Notice that left inverse element a'_L for element a can be defined only when there is a left identity e_L wrt *. Similarly

Binary Operations and Groups

right inverse element a'_R for a can be defined only when there is a right identity e_R wrt *.

Definition 1.3.4 : Whenever $a'_L = a'_R$ for an element $a \in A$ wrt *, the common value is denoted by a' and is called *inverse element* of element a wrt *.

Thus the element a' is inverse element wrt * of element a if $a' * a = a * a' = e$, where e is identity element wrt *. Notice that inverse element a' for element a can be defined only when there is identity element e wrt *.

1.3.1 Illustrative Examples

1. (*i*) For b.o. + over the set $\mathbb{Z}$, since $0 + m = m$, true for all $m \in \mathbb{Z}$, thus $e_L = 0$ wrt +. Again, $m + 0 = m$, true for all $m \in \mathbb{Z}$ thus $e_R = 0$ wrt + ; and since $e_L = e_R = 0$, therefore $e = 0$ for b.o. + over $\mathbb{Z}$. Similarly for any other set, *identity element is 0 when * = +*. (Does any other number except 0 satisfy the conditions of e_L and e_R ?)

(*ii*) For * = − defined over set $\mathbb{R}$, since $0 - x = -x$, true for all $x \in \mathbb{R}$, thus $e_L \neq 0$ wrt * = −. Similarly no other number will satisfy $e_L * a = a$, for any a $\in \mathbb{R}$, thus operation - defined over $\mathbb{R}$ (or any other set) does not have left identity element. Since $x - 0 = x$, true for all $x \in \mathbb{R}$, thus $e_R = 0$ wrt * = − · (Do you have any number other than 0 satisfying the condition of e_R ?). So, when * = − there is no left identity element and 0 is right identity element and hence *identity element e does not exist for * = −*.

(*iii*) For * = · defined over $\mathbb{R}$, $e_L = 1$ as $1 \cdot x = x$ for all $x \in \mathbb{R}$ and $e_R = 1$ as $x \cdot 1 = x$ for all $x \in \mathbb{R}$. Thus $e_L = e_R = 1$ so that *for multiplication identity element e is 1*.

(*iv*) For the operation * = ÷ defined over $\mathbb{R}$, identity element e_L does not exist as no number satisfies the condition $e_L \div x = x$ for all $x \in \mathbb{R}$ while $e_R = 1$, as $x \div 1 = x$ for all $x \in \mathbb{R}$. So, when * = ÷ there is no left identity element and 1 is right identity element and hence identity element **e** does not exist for * = ÷.

(v) For the generalized operation $*$ defined over $\mathbb{R}$ by $a * b = |a| b$, $e_L = -1$ and 1.

Since no number satisfying $a * e_R = a$ exists, thus e_R does not exist for $*$ defined above.

(vi) For the generalised operation $*$ defined over $\mathbb{Q}$ by $p * q = pq + 1$ for $p, q \in \mathbb{Q}$, let us find out e_L and e_R.

Since $e_L * q = q$ which by given rule $*$ implies $e_L q + 1 = q \Rightarrow e_L = \dfrac{q-1}{q}$ provided $q \neq 0$. similarly $q * e_R = q$

$\Rightarrow q e_R + 1 = q \Rightarrow e_R = \dfrac{q-1}{q}$, $q \neq 0$. Thus

$e_L = e_R$ and hence $e = \dfrac{q-1}{q}$ for the operation $*$ defined above where $q \in \mathbb{Q}^*$.

2. We have seen in above examples that when $* = +$, $e = 0$ and when $* = \cdot$, $e = 1$. Thus we can obtain inverse elements for these two operations. When $* = -$ or $\div$, e does not exist and hence inverse elements wrt $-$ and $\div$ can not be obtained.

Since $(-m) + m = m + (-m) = 0$, the e of $* = +$, thus $a'_L = a'_R = -m$ for $a = m$. Thus when operation is $+$, inverse element of an element x will be its negative $-x$. Similarly, since $\dfrac{1}{m} \cdot m = m \cdot \dfrac{1}{m} = 1$, the identity e of $* = \cdot$, thus $a'_L = a'_R = \dfrac{1}{m}$ for $a = m$. Thus when operation is $\cdot$, inverse element of an element x will be its reciprocal $\dfrac{1}{x}$.

3. For the operation $*$ defined over $\mathbb{R}$ by $a * b = |a| b$, right inverse element for any element can not exist as e_R does not exist for $*$; and hence inverse element can not exist for $*$ defined above.

4. For the operation $*$ defined over $\mathbb{Q}$ by $p * q = pq + 1$, let us find out inverse elements.

Since $a'_L * q = e_L = \dfrac{q-1}{q}$, $q \neq 0$ (by example 1 (vi)),

$$a'_L q + 1 = \dfrac{q-1}{q}$$

$\Rightarrow a'_L = -\dfrac{1}{q^2}$, for any element $q \neq 0$.

Similarly $q * a'_R = e_R = \dfrac{q-1}{q}$, $q \neq 0 \Rightarrow qa'_R + 1$

$$= \dfrac{q-1}{q}$$

$\Rightarrow \qquad a'_R = -\dfrac{1}{q^2}$

Thus $\qquad a'_L = a'_R = -\dfrac{1}{q^2}$ when $a = q \neq 0$.

1.4 Groups

We are now in a position to define a very important concept of modern abstract algebra - the groups. The definition of a group given here as definition 1.4.5 is the *axiomatic definition* of a group and was formulated in 1882 by Heinrich Weber.

Definition 1.4.1 : A system containing a non-empty set L, one or more binary operations defined over L together with a set of specific conditions imposed on the binary operations (called the *axioms*) is said to be an *algebraic structure*.

Groups are one of the many algebraic structures. Other algebraic structures include *Rings, Fields, Integral Domains* and *Vector spaces*; all of which play significant roles in the theory and applications of modern algebra.

Definition 1.4.2 : The pair of a non-empty set G and a b.o. * defined over G is called a *Groupoid*, denoted by the symbol (G, *). If the operation * is binary only for a few elements of G then the pair (G, *) is called a *half-groupoid*.

Example : The pair ($\mathbb{Z}^+$, +) of the set of positive integers and b.o. + over $\mathbb{Z}^+$ is an example of a groupoid. The pair ($\mathbb{Z}^+$, –) is half groupoid because the operation * = – is not binary for all the elements of $\mathbb{Z}^+$.

Definition 1.4.3 : The pair (G, *) of a non-empty set G and a b.o. * defined over G is called a *Semi group* if * satisfies associative law as well.

Thus a semi group is a groupoid in which the b.o. * is associative.

Example : The groupoids ($\mathbb{Z}^+$, +), ($\mathbb{Z}^+$, ·) are examples of a semi group while the groupoid ($\mathbb{Z}$, –) is not a semi group.

Definition 1.4.4 : The pair (G, *) of a non-empty set and b.o. * defined over G is said to be a *Monoid* if * is associative and G contains the identity element of *.

Thus a monoid is a semigroup which contains the identity element of * in it.

Example : The semi group ($\mathbb{Z}^+$, ·) is a monoid but the semi group ($\mathbb{Z}^+$, +) is not a monoid. (Why ?)

Definition 1.4.5 (Definition of a group) : The pair (G, *) of a non-empty set G and a b.o. * defined over G is said to be a group if * satisfies the following conditions :

(G1)   * is associative in G, *i.e.* for any three elements a, b, c of G

$$a * (b * c) = (a * b) * c$$

(G2)   G contains the identity element of *. Thus if e is identity element of *, *i.e.* if $e * a = a * e = a$ for all elements a of G then $e \in$ G.

(G3)   G contains inverse of each of its elements. Thus if a' is inverse element of any element $a \in$ G wrt *, *i.e.* if $a' * a = a * a' = e$ where e is identity element of * then $a' \in$ G.

If, in addition to above three conditions, ∗ is commutative as well, i.e.

(G4) $a * b = b * a$ for any two elements a, b of G then the group (G, ∗) is said to be an *abelian** *group* or *a commutative group*.

Note : 1. The above conditions (G1) – (G3) are called the *group-axioms*. The axiom (G4) is additional condition for abelian groups.

2. The condition of ∗ being a b.o. over the set G can be restated by saying that *operation* ∗ *is closed in* G.

3. The axiom (G2) implies that $e_L = e_R = e$ and $e \in G$ and the axiom (G3) implies that $a'_L = a'_R = a'$ and $a' \in G$ for any element $a \in G$.

4. In view of (3) above, either of the conditions $e * a = a$, $a * e = a$ of axiom (G2) and $a' * a = e$, $a * a' = e$ of axiom (G3) can be used for the purpose of testing the axioms (G2) and (G3) in any pair (G, ∗).

5. Most of the times we denote a group (G, ∗) simply by G with the understanding that group G will always have a b.o. ∗ defined over G.

Definition 1.4.6 (Order of a group) : The number of elements in the set G is the order of G, denoted by 0 (G).

If the set G is finite, group G is said to be a *finite group* and if the set G is infinite, group G is said to be an *infinite group*.

---

* Niels Heinrich Abel : A commutative group was named an abelian group in the honour of Niels Heinrich Abel whose main interest was in the group of permutations, especially groups of even permutations. Born on August 5, 1802 in Finnoy, Norway, Abel (1802-1829) contributed significantly in the theory of elliptic functions. He applied group theoretical concepts to the theory of equations along with Evariste Galois (1811-1832), the inventor of Galois theory on groups, Abel and Galois both died very young.

Thus for a finite group, $0(G)$ is finite and for an infinite group $0(G) = \infty$.

1.4.1. Illustrative Examples

(A) Groups of Numbers

1. The smallest possible group is the one-element group $(\{e\}, *)$ where e is the identity element of the operation $*$, defined by $e * e = e$. Thus the pairs $(\{0\}, +), (\{1\}, \cdot)$ are examples of one-element groups.

2. The set $\{-1, 1, i, -i\}$ (which infact is the set of fourth roots of unity) forms a finite abelian group wrt $* = \cdot$ (multiplication), which is a b.o. over it. Here $e = 1$, $(-1)' =$ inverse of $-1 = -1$, $1' = 1$, $i' = -i$ and $(-i)' = i$. Elements of the set satisfy associative and commutative laws wrt $* = \cdot$ and the group is finite as $0(G) = 4^{\dagger}$ Does set of cube roots of unity $\{1, \omega, \omega^2 : \omega^3 = 1\}$ also form a finite abelian group when $* = \cdot$?

3. The pairs $(\mathbb{Z}^+, +), (\mathbb{Z}^+ \cup \{0\}, +), (\mathbb{Z}, -), (\mathbb{Z}, \cdot), (\mathbb{R}, \cdot)$ and $(\mathbb{R}^*, \div)$ do not form groups. The respective reasons are :

   (i) The identity element 0 for the operation + does not belong to $\mathbb{Z}^+$, violating the condition (G2). Hence $(\mathbb{Z}^+, +)$ is not a group.

   (ii) The inverse element $-m$ for any element $m \in \mathbb{Z}^+ \cup \{0\}$ for the operation + does not belong to the set violating the group axiom (G3) and hence $(\mathbb{Z}^+ \cup \{0\}, +)$ is not a group.

   (iii) The operation − is not associative over $\mathbb{Z}$ violating the group axiom (G1), hence $(\mathbb{Z}, -)$ is not a group. Secondly, for operation − only e_R exists and not e_L. Due to same reasons $(\mathbb{R}, -)$ is also not a group. In fact, no set can form a group with − (usual subtraction).

---

†Similarly the set $\{-1, 1\}$ also forms a group with operation $* = \cdot$ and the group is finite and abelian.

(iv) The inverse element $\frac{1}{m}$ for any element $m \in \mathbb{Z}$ for the operation · (multiplication) does not belong to the set, hence $(\mathbb{Z}, \cdot)$ is not a group.

(v) The pair $(\mathbb{R}, \cdot)$ would have formed a group but for the element $0 \in \mathbb{R}$, whose inverse $\frac{1}{0}$ with respect to operation · does not belong to $\mathbb{R}$. Hence $(\mathbb{R}^*, \cdot)$ forms an infinite group which is abelian as well. Similarly $(\mathbb{Q}^+, \cdot)$, $(\mathbb{Q}^*, \cdot)$ are also infinite abelian groups. $(\mathbb{Q}^-, \cdot)$, $(\mathbb{R}^-, \cdot)$ are not groups, why ?

(vi) For the operation ÷ only e_R exists (which is 1). There is no e_L for ÷. Thus no set can form a group with operation ÷ (usual division).

4. *Groups of integers* : One of the most commonly used groups is the group $(\mathbb{Z}, +)$, called the group of integers. Operation + which is a b.o. over $\mathbb{Z}$ satisfies associative law with the elements of $\mathbb{Z}$, its identity element $0 \in \mathbb{Z}$ and each element m of $\mathbb{Z}$ has its inverse $-m$ present in $\mathbb{Z}$. Also + satisfies commutative law, hence $(\mathbb{Z}, +)$ is an abelian group whose order is infinite.

Similarly for a fixed integer n the pair $(n\mathbb{Z}, +)$ and the pairs $(\mathbb{Q}, +)$, $(\mathbb{R}, +)$ all form infinite abelian groups. Thus the set of all even integers $2\mathbb{Z}$ forms a group with +. Does the set of all odd integers also form a group with + ?

5. *Groups of complex numbers* : (i) The set $\mathbb{C}$ of all complex numbers forms a group with * taken as addition of two complex numbers. The group is infinite and abelian. The operation + of addition of two complex numbers is a b.o. over $\mathbb{C}$ and satisfies associative law; its identity element $0 \in \mathbb{C}$ and each element $a + ib$, where $a, b \in \mathbb{R}$, of $\mathbb{C}$ has its inverse $(-a) + i(-b)$ present in $\mathbb{C}$. Also for any two complex numbers z_1 and z_2, $z_1 + z_2 = z_2 + z_1$.

(ii) The set $\mathbb{C}^*$ of all non-zero complex numbers forms a group with the operation taken as multiplication of two complex numbers given by

$$(a + ib)(c + id) = (ac - bd) + i(ad + bc)$$

Since $(ac - bd) \neq 0$ and $(ad + bc) \neq 0$ as $a + ib \neq 0$ and $c + id \neq 0$, hence the product $(a + ib)(c + id)$ is a non-zero complex number so that operation is closed. Multiplication of complex numbers is associative and the element $1 + i0$ serves as identity element of the operation and is present in $\mathbb{C}^*$. Finally inverse of $(a + ib)$ wrt $*$ is $\frac{1}{a + ib}$ and since $\frac{1}{a + ib} = \frac{a}{a^2 + b^2} + i\left(\frac{-b}{a^2 + b^2}\right)$, thus the inverse of $a + ib$ is present in $\mathbb{C}^*$.

This group is infinite abelian group because $0(\mathbb{C}^*) = \infty$ and complex numbers obey commutative law with multiplication as the operation.

The pair $(\mathbb{C}, \cdot)$ is not a group, why?

(iii) The set of all complex numbers $\mathbb{C}$ does not form a group with multiplication of complex numbers as a b.o. but the set of all complex numbers z satisfying $|z| = 1$, i.e. the set (the unit circle) $O_1 = \{z \in \mathbb{C} : |z| = 1\}$ forms a group under this operation.

If $z_1, z_2 \in \mathbb{C}$ such that $|z_1| = 1$, $|z_2| = 1$ then since $|z_1 z_2| = |z_1| |z_2| = 1$, operation is closed. From (ii) above $e = 1 + i0$ and since $|1 + i0| = 1$, identity element is present in the set O_1.

Also inverse $(a + ib) = \frac{1}{a + ib} = \frac{a}{a^2 + b^2} + i\left(\frac{-b}{a^2 + b^2}\right)$

where $\left|\frac{a}{a^2 + b^2} + i\left(\frac{-b}{a^2 + b^2}\right)\right| = \sqrt{\frac{a^2}{(a^2 + b^2)^2} + \frac{b^2}{(a^2 + b^2)^2}}$

$$= \frac{1}{\sqrt{a^2 + b^2}} = 1$$

hence inverse of $(a + ib)$ is also present in the set O_1. Associativity of elements of O_1 follows from properties of complex numbers. Hence $(O_1, \cdot)$ is a group and this group is called the *circle group*.

6. *Groups of n-tuples of numbers* : Consider the set $\mathbb{R}^n$ of all n-tuples of real numbers, $\mathbb{R}^n = \{x = (x_1, x_2, ..., x_n) \text{ where } x_1, x_2, ..., x_n \in \mathbb{R}\}$, taking $*$ as co ordinate wise addition in $\mathbb{R}^n$, given by

$$x + y = (x_1, x_2, ..., x_n) + (y_1, y_2, ..., y_n)$$
$$= (x_1 + y_1, x_2 + y_2, ..., x_n + y_n)$$

then $(\mathbb{R}^n, *)$ forms an abelian group. If $x, y \in \mathbb{R}^n$ then $x + y$ defined above also belongs to $\mathbb{R}^n$ so that operation is closed. Here $e = (0, 0, ..., 0) \in \mathbb{R}^n$, inv of $x = x' = (-x_1, -x_2, ..., -x_n) \in \mathbb{R}^n$ and also $x + y = y + x$. Associativity of elements of $\mathbb{R}^n$ follows from the properties of real numbers, *i.e.* $x + (y + z) = (x + y) + z$, $x, y, z \in \mathbb{R}^n$.

Similarly the set $\mathbb{C}^n$ of all n-tuples of complex numbers forms an abelian group with co ordinate wise addition taken as the b.o.

7. *Groups of numbers with generalized operations* : A generalized operation is a combination of two or more arithmetical operations + (addition), − (subtraction), · (multiplication) and ÷ (division). Let us consider some pairs of sets of numbers and certain generalized operations.

(*i*) The set $\mathbb{Q}$ of all rational numbers with b.o. defined by $a * b = a \cdot b + 1$, $\forall\, a, b \in \mathbb{Q}$ does not form a group because the given operation $*$ is not associative (see Example 3 of 1.2.3).

Consider the set $\mathbb{Q}^+$ of all positive rational numbers and the operation $*$ over it defined by $a * b = \frac{1}{2} a \cdot b\, \forall\, a, b \in \mathbb{Q}^+$. Since $\frac{1}{2} a \cdot b \in \mathbb{Q}^+$, operation is binary. For any $p, q, r \in \mathbb{Q}^+$, you can check that $p * (q * r) = \frac{1}{4} prq = (p * q) * r$ and hence given $*$ satisfies associative law. By definition of e_L,

$e_L * p = p$ or $\frac{1}{2} e_L \cdot p = p \Rightarrow e_L = 2$. Similarly $p * e_R = p$ or $\frac{1}{2} p \cdot e_R = p \Rightarrow e_R = 2$ so that $e_L = e_R = e = 2$, which is present in $\mathbb{Q}^+$. By definition of p'_L,

$$p'_L * p = e \text{ or } \frac{1}{2} p'_L \cdot p = 2 \Rightarrow p'_L = \frac{4}{p}.$$

Similarly $p'_R = \frac{4}{p}$ so that p' (inv. of p) $= \frac{4}{p}$ which belongs to $\mathbb{Q}^+$. Thus the pair $(\mathbb{Q}^+, *)$ forms a group. This group is abelian as well because for any $p, q \in \mathbb{Q}^+$, $p * q = \frac{1}{2} p \cdot q = \frac{1}{2} q \cdot p = q * p$.

(ii) The set $\mathbb{R} - \{-2\}$ of all real numbers except the element -2 forms an infinite abelian group wrt operation $*$ defined by

$$a * b = 2a + 2b + ab + 2, \forall\, a, b \in \mathbb{R} - \{-2\}.$$

Since $2a + 2b + ab + 2 \in \mathbb{R} - \{-2\}$, operation $*$ given here is a b.o. Also for any $a, b, c \in \mathbb{R} - \{-2\}$, $a * (b * c) = (a * b) * c$ (Find the common value).

Assuming e to be existing for $*$, $e * a = a$ implies $2e + 2a + ea + 2 = a \Rightarrow e = -1 \in \mathbb{R} = -\{-2\}$, and

$$a' * a = e \Rightarrow 2a' + 2a + a'a + 2 = -1$$

$$\Rightarrow a' = \frac{-(2a + 3)}{2 + a} \in \mathbb{R} - \{-2\}.$$

For example if $a = 4$ then $a' = -11/6 \in \mathbb{R} - \{-2\}$. Also from the given definition

$$a * b = 2a + 2b + ab + 2$$
$$= 2b + 2a + ba + 2$$
$$= b * a,$$

$*$ satisfies commutative law.

Some more groups of numbers with generalized operations have been given in the solved problem set - I.

(B) Groups of Matrices

1. The set $\mu_{m \times n}(\mathbb{R})$ of all $m \times n$ matrices having their elements from the set of real numbers $\mathbb{R}$ forms an infinite abelian group

Binary Operations and Groups

with the b.o. * taken as addition of matrices. An element A of the set $\mu_{m \times n}(\mathbb{R})$ is of the form

$$A = \begin{bmatrix} a_{11} & a_{12} & \cdots & a_{1n} \\ a_{21} & a_{22} & \cdots & a_{2n} \\ \vdots & & & \\ a_{m1} & a_{m2} & \cdots & a_{mn} \end{bmatrix}_{m \times n}$$

where each a_{ij} ($i = 1, 2, \ldots, m; j = 1, 2, \ldots, n$) is a real number.

If A, B, $\in \mu_{m \times n}(\mathbb{R})$ then A + B and $-$A also belong to $\mu_{m \times n}(\mathbb{R})$. The null matrix 0 of order $m \times n$ will be the identity element and inv of A, $A^1 = -A$. Associativity and commutativity of the elements follows from the properties of $m \times n$ matrices, because for any A, B, C $\in \mu_{m \times n}(\mathbb{R})$ we know that A + (B + C) = (A + B) + C and A + B = B + A. Thus $(\mu_{m \times n}(\mathbb{R}), +)$ is an infinite abelian group.

Note : In the above group, the elements of matrices may also be from the sets of integers, rationals or complex numbers, and the corresponding groups are denoted by $(\mu_{m \times n}(\mathbb{Z}), +)$, $(\mu_{m \times n}(\mathbb{Q}), +)$ and $(\mu_{m \times n}(\mathbb{C}), +)$ respectively.

2. The set $v_{n \times n}(\mathbb{R})$ of all $n \times n$ non-singular square matrices (so that elements of $v_{n \times n}(\mathbb{R})$ have inverses) having their elements from the set of real numbers $\mathbb{R}$ forms an infinite non-abelian group wrt the b.o. * taken as multiplication of matrices. An element A of this set is of the form

$$A = \begin{bmatrix} a_{11} & a_{12} & \cdots & a_{1n} \\ a_{21} & a_{21} & \cdots & a_{2n} \\ & & & \\ a_{n1} & a_{n2} & \cdots & a_{nn} \end{bmatrix}_{n \times n},$$

where each a_{ij} ($i = 1, 2, \ldots, n; j = 1, 2, \ldots, n$) is a real number, and determinant of A, $|A|$, is non-zero.

If A, B $\in v_{n \times n}(\mathbb{R})$ then since $|AB| = |A| |B| \neq 0$, AB $\in v_{n \times n}(\mathbb{R})$, the operation is closed. The unit matrix I of order $n \times n$ will be the identity element and inv of A = A' will be the matrix $\frac{\text{Adj } A}{|A|}$; both belonging to $v_{n \times n}$ ($\mathbb{R}$).

Associativity A (BC) = (AB) C for any elements A, B, C $\in$ $v_{n \times n}$ ($\mathbb{R}$) follows from the properties of $n \times n$ square matrices. Since, in general, AB $\neq$ BA, the group ($v_{n \times n}$ ($\mathbb{R}$), $\cdot$) is an infinite non-abelian group.

Note : 1. In the above group, elements of matrices may also be from the sets of rationals or complex numbers but not from the set of integers. In case elements are integers, the inverse A' of such a matrix A may not belong to $v_{n \times n}$ ($\mathbb{Z}$). For example, if $A = \begin{bmatrix} 2 & 1 \\ 3 & 4 \end{bmatrix}$ then since $|A| = 5 \neq 0$, thus $A \in v_{n \times n}$ ($\mathbb{Z}$) but

$$\text{inv } A = A' = \begin{bmatrix} 4/5 & -1/5 \\ -3/5 & 2/5 \end{bmatrix} \notin v_{2 \times 2} (\mathbb{Z}).$$

2. Due to frequent occurrence of groups of $n \times n$ square matrices in topics of linear algebra, we also denote the group ($v_{n \times n}$ ($\mathbb{R}$), $\cdot$) by GL_n ($\mathbb{R}$) and call it *general linear group* over $\mathbb{R}$. Similarly we have general linear groups GL_n ($\mathbb{Q}$) and GL_n ($\mathbb{C}$) over $\mathbb{Q}$ and $\mathbb{C}$ respectively. For $n > 1$, all the general linear groups are infinite and non-abelian. For example, GL_2 ($\mathbb{R}$) denotes the group of all 2×2 non-singular matrices ($v_{n \times n}$ ($\mathbb{R}$), $\cdot$) with their elements from set of real numbers and the b.o. $*$ as matrix multiplication. Thus GL_2 ($\mathbb{R}$) = $\left(\begin{bmatrix} a & b \\ c & d \end{bmatrix} : a, b, c, d \in \mathbb{R} \right.$

and $ad - bc \neq 0$, $* = \cdot \Big)$

The group is infinite as a, b, c, d are any real numbers and non-abelian as, for example,

$$\begin{bmatrix} 1 & 3 \\ 2 & 4 \end{bmatrix} \begin{bmatrix} 6 & -5 \\ 8 & 7 \end{bmatrix} = \begin{bmatrix} 30 & 16 \\ 44 & 18 \end{bmatrix} \neq \begin{bmatrix} 6 & -5 \\ 8 & 7 \end{bmatrix} \begin{bmatrix} 1 & 3 \\ 2 & 4 \end{bmatrix}$$

$$= \begin{bmatrix} -4 & -2 \\ 22 & 52 \end{bmatrix}$$

(C) Groups of Functions

(i) The set B_A of all one-to-one mappings of non-empty set A onto itself forms an infinite non-abelian group with b.o. as the composition of mappings.

If $f_1, f_2 \in B_A$, i.e. if $f_1 : A \to A$ and $f_2 : A \to A$ are two one-to-one maps of set A onto itself, then from Prelim. [3 (vi) and (vii)], we know that their composition (fg) defined by $(fg)(x) = f(g(x))$ for all $x \in A$ is also one-to-one map from A onto A so that the operation is closed and that the composition (fg) satisfies associative law. Since $(If_1) = (f_1 I) = f_1$, where $I : A \to A$ is the identity map, hence identity element of this group is the map I. Inverse of $f_1 \in B_A$ is the inverse map f_1^{-1} because from [Prelim. 3 (iii)],

$$(f_1^{-1} f_1) = (f_1 f_1^{-1}) = I$$

This group is non-abelian as $f_1, f_2 \in B_A$ does not imply $f_1 f_2 = f_2 f_1$ [See Prelim. 3 (vii)]. This group is called the *transformation group of A*.

(ii) Apart from above examples on groups, we also have groups of permutations and groups of symmetries of geometrical figures like rectangles, squares and circles. For these groups, you have to wait till chapter II !

1.5 Group Tables

For small finite groups, we can arrange elements of the group in a tabular form in such a way that all the group axioms can be verified from the table. Such tables are called group tables. As in b.o. tables (see 1.2), in group tables also, the elements of the set forming group are written along the top-most row and along the extreme left corner of the table. But, unlike b.o. tables, in group tables each element of the group (including identity element e) appears only once in each row and in each column. A group table has following two characteristics :

(i) In the group table, the elements below e and the elements against e appear in the same order as given in the set. This helps us in locating e immediately.

(*ii*) If the group is abelian, the group table will be symmetrical with respect to the principal diagonal.

1.5.1 Examples of Group Tables

1. Let the group G has 2 elements $\{e, a\}$. Then its group-table will be of the form.

b.o. *	e	a
e	e	a
a	a	e

Table 1.5.1

There can not be any other group table of a group having two elements. Thus for a group of two elements, $a * a$ has to be e.

2. Let the group G has 3 elements $\{e, a, b\}$. Then its group-table will be of the form :

b.o. *	e	a	b
e	e	a	b
a	a	b†	e
b	b	e	a

Table 1.5.2

(† This place of second row and second column of the table can be filled either by e or by b. If filled by e the next place in the second row wll have to be filled by b and in that case column three will have element b appearing twice in it. Hence the element b at place †.)

There can not be a group table other than table 1.5.2 of a group having 3 elements. Thus in a group of 3 elements $\{e, a, b\}$, $a * a = b$, $b * b = a$ and $a * b = e$. Notice also that the elements against e in the first row and those below e in the first column of table 1.5.2 appear in the order e, a, b; the order of the elements in the set $\{e, a, b\}$.

3. Let the group G has 4 elements $\{e, a, b, c\}$. In this case, two independent group tables can be made. They are :

Binary Operations and Groups

b.o. *	e	a	b	c
e	e	a	b	c
a	a	e	c	b
b	b	c	e	a
c	c	b	a	e

Table 1.5.3 (Klein's group table)

b.o. *	e	a	b	c
e	e	a	b	c
a	a	b	c	e
b	b	c	e	a
c	c	e	a	b

Table 1.5.4

Any other group table of a group having 4 elements will be equivalent to either of these two tables.

Table 1.5.3: The group described by the table 1.5.3 is called *Klein's 4-group* or simply *Klein's group*, after the name of German mathematician Felix Klein*, and is denoted by K_4 or by V (first letter of the German word 'viergruppe'; vier meaning four). Klein's group table has a special significance as most of the groups of 4 elements appearing in various applications follow it. In Klein's group table, $a * a = e$, $b * b = e$, $c * c = e$ and $a * b = c$, $a * c = b$, etc.; *i.e.* when same elements are operated we get e and when two distinct elements are operated we get the third non-identity element of the group.

Table 1.5.4: The elements e, a, b, c in the group table 1.5.4

Felix Klein (1849-1925): Felix Klein contributed mainly on the groups of linear transformations which included reflections and rotations in two and three dimension and investigated the concept of discrete groups. His paper on groups of transformations published jointly with Sophus Lie (1842-1899), another pioneer in the field of abstract algebra, in the year 1871 gave a new dimension to the theory of groups. His Klein's 4-group is significantly important as this group is the only finite abelian and non-cyclic group of rotation in a three dimensional space.

move in a cyclic order. In particular, if the four elements are 0, 1, 2, 3 we get from table 1.5.4.

b.o. * = $+_4$	0	1	2	3
0	0	1	2	3
1	1	2	3	0
2	2	3	0	1
3	3	0	1	2

Table 1.5.5 (Group table of $\mathbb{Z}_4$)

Notice that the four elements in the table 1.5.5 move according to the operation of addition modulo 4 ($+_4$) (See Prelim. 2(v)). This special form of 4 elements group with the elements 0, 1, 2, 3 is denoted by $\mathbb{Z}_4$ (or by $\mathbb{C}_4$). Here b.o. is $+_4$. $\mathbb{Z}_4$-groups is one of the members of family of $\mathbb{Z}_n$-groups where set $\mathbb{Z}_n$ {0, 1, 2, ..., $n-1$} and b.o. * = $+_n$ (addition modulo n). For example, the group table for $\mathbb{Z}_5$-group will be :

b.o. * = $+_5$	0	1	2	3	4
0	0	1	2	3	4
1	1	2	3	4	0
2	2	3	4	0	1
3	3	4	0	1	2
4	4	0	1	2	3

Table 1.5.6. (Group table for the group $\mathbb{Z}_5$)

Note : 1. Notice that all group tables above of the groups of 1 element (make its table yourself !), 2 elements, 3 elements and both the tables of 4 elements are symmetrical with respect to the principal diagonal showing that the elements of groups of order 1, 2, 3 and 4 obey commutative law and hence all the groups of orders upto 4 are abelian groups.

2. The group tables 1.5.5 and 1.5.6 of the groups $\mathbb{Z}_4$ and $\mathbb{Z}_5$ are symmetrical about the principal diagonal and hence they are abelian. By making some more group tables of $\mathbb{Z}_n$-groups, you will see that all $\mathbb{Z}_n$-groups are abelian groups.

1.6 Elementary Properties of a Group

In the following theorems we have arranged some of the common properties of a group obtainable from group axioms.

Theorem 1.6.1 : In a group (G, $*$), left and right cancellation laws hold true.

Thus in a group (G, $*$) we have

(i) $a * b = a * c \Rightarrow b = c$ (left cancellation law (in short LCL))

and (ii) $a * c = b * c \Rightarrow a = b$ (right cancellation law (in short RCL))

where $a, b, c \in G$.

Proof : (i) Since $a \in G, a' \in G$. Pre-operate given equation

$$a * b = a * c \quad \text{by } a', \text{ we get}$$
$$a' * (a * b) = a' * (a * c)$$
$\Rightarrow \quad (a' * a) * b = (a' * a) * c \quad \text{(by AL)}$
$\Rightarrow \quad e * b = e * c \quad \text{(definition of } e\text{)}$
$\Rightarrow \quad b = c$

(ii) Post operate by c' (inv. of c) in the given equation and proceed exactly as in (i).

Corollary 1.6.1 : In a group (G, $*$), the left identity e_L is also the right identity e_R.

Proof : Consider $a'_L * (a * e_L)$

$= (a'_L * a) * e_L$

$= e_L * e_L$ (by definition of left inverse e_L)

$= e_L$

$= a'_L * a$

$\Rightarrow \quad a'_L * (a * e_L) = a'_L * a$

$\Rightarrow \quad a * e_L = a \quad \text{(by LCL)},$

showing that e_L is the right identity as well. Alternatively, if e_L is the left identity then $e_L * e_R = e_R$ and if e_R is the right identity then $e_L * e_R = e_L$ giving us $e_R = e_L$.

Corollary 1.6.2 : In a group (G, *), the left inverse a'_L of an element a is also the right inverse a'_R.

Proof : Consider $a'_L * (a * a'_L)$
$$= (a'_L * a) * a'_L$$
$$= e_L * a'_L$$
$$= a'_L * e_L$$
$$= a'_L * e_R \quad \text{(as } e_L \text{ is also the right identity)}$$
$\Rightarrow a'_L * (a * a'_L) = a'_L * e_R$
$\Rightarrow \qquad a * a'_L = e_R,$

Showing that a'_L is the right inverse as well.

Theorem 1.6.2 : In a group (G, *) identity element is unique. (In fact, identity element is unique for any operation *).

Proof : Let if possible there be two identity elements e_1 and e_2 in (G, *). Then for any $a \in G$, $a * e_1 = a$ and $a * e_2 = a$. Thus $a * e_1 = a * e_2$ which by LCL implies $e_1 = e_2$.

Theorem 1.6.3 : In a group (G, *), inverse element of any element a is unique.

Proof : Let, if possible, there be two inverse elements a' and a'' for element a in G. Then $a * a' = e$ and $a * a'' = e$. Thus $a * a' = a * a''$ which by LCL implies $a' = a''$.

Theorem 1.6.4 : In a group (G, *) the linear equations

$$a * x = b \qquad \ldots(i)$$
and
$$y * a = b \qquad \ldots(ii)$$

where $a, b \in G$, have unique solutions in G.

Proof : Consider (i). Pre operate on both sides by a', we get $a' * (a * x) = a' * b$
$\Rightarrow \qquad (a' * a) * x = a' * b \quad \text{(by AL)}$
$\Rightarrow \qquad e * x = a' * b$
$\Rightarrow \qquad x = a' * b$

Hence $x = a' * b$ is a solution of (i). Since $a', b \in G$, thus $x = a' * b$ also belongs to G.

For uniqueness of solution, assume that (i) has two solutions x_1 and x_2 in G. then $a * x_1 = b$ and $a * x_2 = b$, which implies, $a * x_1 = a * x_2$ or $x_1 = x_2$. Thus solution is unique.

Proceed exactly in case of equation (ii) and show that $y = b * a'$ is an unique solution of (ii).

Theorem 1.6.5 : In a group (G, *), $(a * b)^{-1} = b^{-1} * a^{-1}$, where a^{-1} is inv of a and b^{-1} is inv of b and $a, b \in G$.

Proof : In order to prove that

$x^{-1} = k$, we need to prove that $x * k = e$ as well as $k * x = e$. Thus consider

$$\begin{aligned}
& (a * b) * (b^{-1} * a^{-1}) \\
= & (a * b) * k \quad (k = b^{-1} * a^{-1}) \\
= & a * (b * k) \quad \text{(by AL)} \\
= & a * (b * (b^{-1} * a^{-1})) \\
= & a * ((b * b^{-1}) * a^{-1}) \quad \text{(by AL)} \\
= & a * (e * a^{-1}) \\
= & a * a^{-1} \\
= & e
\end{aligned}$$

Again, $(b^{-1} * a^{-1}) * (a * b)$

$$\begin{aligned}
= & k * (a * b) \\
= & (k * a) * b \\
= & ((b^{-1} * a^{-1}) * a) * b \\
= & (b^{-1} * (a^{-1} * a)) * b \\
= & (b^{-1} * e) * b \\
= & b^{-1} * b \\
= & e
\end{aligned}$$

Thus $\quad (a * b)^{-1} = b^{-1} * a^{-1}$

This is the *reversal rule* of inverses. In fact, by its repeated application, it can be generalized as

$(a_1 * a_2 * \ldots * a_n)^{-1} = a_n^{-1} * \ldots * a_2^{-1} * a_1^{-1}$, where $a_1, a_2, ..., a_n \in G$.

Theorem 1.6.6 : If in a group $(G, *)$, $a^2 = e$ where a^2 means $a * a$ for any $a \in G$ then G is an abelian group.

Proof : Let $a, b \in G$. Then $a * b \in G$ and as given, $a^2 = e$, $b^2 = e$, $(a * b)^2 = e$.

Now $a^2 = e \Rightarrow a * a = e \Rightarrow a = a^{-1}$, where a^{-1} is inv of a. $b^2 = e \Rightarrow b * b = e \Rightarrow b = b^{-1}$, where b^{-1} is inv of b, and

$$(a * b)^2 = e \Rightarrow (a * b) * (a * b) = e$$
$$\Rightarrow (a * b) = (a * b)^{-1}$$
$$= b^{-1} * a^{-1}$$
$$= b * a$$

Thus $a * b = b * a$ or, group G is abelian.

Note : Since $a^2 = e$ implies $a = a^{-1}$, thus in G every element is its own inverse, and we can restate theorem 1.6.6 as

Theorem 1.6.6. (alternative statement) : If in a group $(G, *)$, every element is its own inverse then G is an abelian group.

Theorem 1.6.7 : A group $(G, *)$ is abelian if and only if $(a * b)^2 = a^2 * b^2$, where a^2 means $a * a$ and $a, b \in G$.

Proof : For simplicity of notation, we will drop $*$ from $a * b$ and simply write it as ab with the understanding that between two elements a and b, operation $*$ will always be there. (When we write ab for $a * b$, we say that we are using juxtaposition).

(i) *First part* : Let $(a * b)^2 = a^2 * b^2$, or using juxtaposition

$$(ab)^2 = a^2b^2$$
$$\Rightarrow (ab)(ab) = (aa)(bb)$$
$$\Rightarrow (ab) k_1 = (aa) k_2 \text{ (writing } k_1 \text{ for } ab \text{ and } k_2 \text{ for } bb)$$
$$\Rightarrow a(bk_1) = a(ak_2)$$
$$\Rightarrow bk_1 = ak_2 \quad \text{(by LCL)}$$
$$\Rightarrow b(ab) = a(bb)$$
$$\Rightarrow (ba)b = (ab)b$$
$$\Rightarrow ba = ab \quad \text{(by RCL)}$$

On group G is abelian.

(ii) *Second part* : Let G be abelian so that $ab = ba$.

Binary Operations and Groups

Then $(ab)^2$
$= (ab)(ab)$
$= (ab)(ba)$
$= (ab) k_3$ (writing k_3 for ba)
$= a(bk_3)$ (by AL)
$= a(b(ba))$
$= a((bb)a)$
$= a(b^2a)$
$= a(ab^2)$
$= (aa)b^2$
$= a^2b^2$

Thus $(ab)^2 = a^2b^2$

In general when $n > 0$, $ab = ba \Rightarrow (ab)^n = a^nb^n$.

Theorem 1.6.8: In a group $(G, *)$, $(a^{-1})^{-1} = a$ where a^{-1} is inv of a and $a \in G$.

Proof: Since $a^{-1} * a = e$

Pre-operating both sides by $(a^{-1})^{-1}$ we get

$(a^{-1})^{-1} * (a^{-1} * a) = (a^{-1})^{-1} * e$
$\Rightarrow ((a^{-1})^{-1} * a^{-1}) * a = (a^{-1})^{-1}$ (by AL)
$\Rightarrow e * a = (a^{-1})^{-1}$
$\Rightarrow a = (a^{-1})^{-1}$.

Theorem 1.6.9: A group of order 4 is abelian.

Proof: Consider the group $(G, *)$ where $G = \{e, a, b, c\}$.

(i) If every element in G is its own inverse, then by Theorem 1.6.6, G is abelian.

(ii) If not so then since inverses belong to the group, inverse of one of the 3 elements a, b, c must be the element itself. Let $a^{-1} = a$ so that $a * a = e$, and then $b^{-1} = c$, $c^{-1} = b$ implying $b * c = e$ and $c * b = e$, i.e. $b * c = c * b$.

Now $a * b \neq a$ as $a * b = a \Rightarrow b = e$ which is not true. $a * b \neq b$ as $a * b = b \Rightarrow a = e$ which is not true. Finally $a * b \neq e$ as

$a * b = e \Rightarrow a = b^{-1}$ or $b = a^{-1}$ which is also not true as a and b are not inverses of each other. Thus $a * b = c$. Likewise $b * a = c$ so that $a * b = b * a$. By similar arguments

$a * c = c * a = b$.

Thus a group of 4 elements is necessarily an abelian group.

Solved Problem Set–I

Problem 1.1: How many binary operations can be defined over the set of fourth roots of unity? Of these, how many will be commutative? Give examples of two binary operations over this set.

Solution : The set of fourth roots of unity is $\{1, -1, i - i : i^2 = -1\}$ so that its order is 4.

Thus total number of binary operations that can be defined over this set will be 4^{16}, out of which commutative binary operations will be 4^{10}.

$* = \cdot$ (multiplication) and $* = \div$ (division) are examples of two binary operations over this set, the later being non-commutative b.o. Their b.o. tables are as follows:

$* = \cdot$	1	−1	i	$-i$
1	1	−1	i	$-i$
$-i$	−1	1	$-i$	i
i	i	$-i$	−1	1
$-i$	$-i$	i	1	−1

Commutative b.o. table

$* = \div$	1	−1	i	$-i$
1	1	−1	$-i$	i
−1	−1	1	i	$-i$
i	i	$-i$	1	−1
$-i$	$-i$	i	−1	1

Non-commutative b.o. table

Problem 1.2: Find the left identity e_L, right identity e_R and the identity e in the above two b.o. tables.

Solution : (*i*) When $* = \cdot$, we have from its table,

$1 * a = a$, for all elements a of the set

$\Rightarrow e_L = 1$, and

$a * 1 = a$, for all elements a of the set

$\Rightarrow e_R = 1$. Thus $e_L = e_R = 1 = e$.

(Since $e_L = 1$, the elements in the first row of the table (*i.e.* elements against 1) appear in the order in which they appear in the set. Similarly since $e_R = 1$, the elements in the first column of the table (*i.e.* elements below 1) appear in the order in which they appear in the set.)

(*ii*) when $* = \div$, we have from its table,

$a * 1 = a$ for all elements a of the set

$\Rightarrow e_R = 1$

But there does not exist any element satisfying $e_L * a = a$, for all $a \Rightarrow e_L$ does not exist. Hence e does not exist. (Notice the sequence of elements in column one of the table.)

Problem 1.3 : Show that $(n\mathbb{Z}, +)$ is an abelian group where the set

$$n\mathbb{Z} = \{\ldots -3n, -2n, -n, 0, n, 2n, 3n, \ldots\}$$

is the set of multiples of integers by a fixed integer n.

Solution : (*i*) Let $x, y \in n\mathbb{Z}$ then

$x = rn$ and $y = sn$ for some integers r and s, and $x + y = rn + sn = (r + s)n$

Since $(r + s)$ is an integer, $(r + s) n \in n\mathbb{Z}$ and hence $x + y \in n\mathbb{Z}$ showing that operation $+$ is closed in $n\mathbb{Z}$.

(*ii*) The associativity of elements of $n\mathbb{Z}$ follows from the properties of set of integers $\mathbb{Z}$ as the elements of $n\mathbb{Z}$ are integers.

(*iii*) For $x, y \in n\mathbb{Z}$, since $0 + x = x$ and $x + 0 = x$, $e_L = e_R = 0 = e$.

(*iv*) For $x \in n\mathbb{Z}$, since $x = rn$ for some integer r, there exists $(-r) n$ in $n\mathbb{Z}$ such that $(-r) n + rn = 0$ and also $rn + (-r)n = 0$, so that $x'_L = x'_R = (-r) n = x'$ where $x = rn \in n\mathbb{Z}$. Thus each element in $n\mathbb{Z}$ has its inverse in $n\mathbb{Z}$.

(v) Since $x + y = rn + sn = (r + s)n = (s + r)n$
$= sn + rn = y + x$ $(r, s \in \mathbb{Z})$

$\Rightarrow$ Elements of $n\mathbb{Z}$ satisfy commutative law.

Thus $(n\mathbb{Z}, +)$ is an abelian group whose order is infinite. Particular examples of this group are the groups $(\mathbb{Z}, +)$, $(-\mathbb{Z}, +)$, $(2\mathbb{Z}, +)$, $(-2\mathbb{Z}, +)$, ... When $n = 0$, the group is one element group $(\{0\}, +)$.

Problem 1.4 : Show that following pairs form infinite abelian groups :

(a) $(\mathbb{Q} - \{1\}, *)$ where $*$ is defined as
$a * b = a + b - ab$, for all $a, b \in \mathbb{Q} - \{1\}$

(b) $(\mathbb{Q} - \{-1\}, *)$ where $*$ is defined as
$a * b = a + b + ab$, for all $a, b \in \mathbb{Q} - \{-1\}$.

(c) $(\mathbb{Z}, *)$ where $*$ is defined as
$a * b = a + b + 1$, for all $a, b \in \mathbb{Z}$

(d) $(\mathbb{Q}^+, *)$ where $*$ is defined as
$a * b = \dfrac{ab}{2}$, for all $a, b \in \mathbb{Q}^+$.

Solution : (a) (i) Let $x, y \in \mathbb{Q} - \{1\}$. Thus x is a rational number other than 1 and y is a rational number other than 1. Then $x + y - xy$ can not be equal to 1, for, $x + y - xy = 1 \Rightarrow x + y - xy - 1 = 0 = y(x-1)(1-y) = 0 \Rightarrow x = 1, y = 1$, which is not true. So $x + y - xy \in \mathbb{Q} - \{1\}$ whenever $x, y \in \mathbb{Q} - \{1\}$ and hence

$x * y = x + y - xy \in \mathbb{Q} - \{1\}$ when

$x, y \in \mathbb{Q} - \{1\}$, *i.e.* operation $*$ as defined is a b.o. over the set $\mathbb{Q} - \{1\}$.

(ii) For any 3 elements $l, m, n \in \mathbb{Q} - \{1\}$, we have by definition of $*$:

$l * (m * n) = l * (m + n - mn)$
$= l + (m + n - mn) - l(m + n - mn)$
$= l + m + n - mn - lm - ln + lmn$

Binary Operations and Groups

and
$$(l * m) * n = (l + m - lm) * n$$
$$= (l + m - lm) + n - (l + m - lm)n$$
$$= l + m + n - lm - ln - mn + lmn$$

so that $l * (m * n) = (l * m) * n$

i.e. elements of $\mathbb{Q} - \{1\}$ satisfy associative law for the given operation $*$.

(iii) For any $x \in \mathbb{Q} - \{1\}$, since
$$0 * x = 0 + x - 0 \cdot x = x \Rightarrow e_L = 0$$
and
$$x * 0 = x + 0 - x \cdot 0 = x \Rightarrow e_R = 0$$

Thus for given operation $*$, $e_L = e_R = 0 = e$ and since $0 \in \mathbb{Q} - \{1\}$, the identity belongs to the set.

(iv) By definition, for any $x \in \mathbb{Q} - \{1\}$,
$$x'_L * x = e_L$$
$$\Rightarrow x'_L + x - x'_L x = 0$$
$$\Rightarrow x'_L (1 - x) = -x \Rightarrow x'_L = -\frac{x}{1-x}$$

Similarly $x * x'_R = e_R$
$$\Rightarrow x + x'_R - x x'_R = 0$$
$$x'_R (1 - x) = -x \Rightarrow x'_R = -\frac{x}{1-x}$$

Thus $x'_L = x'_R = -\frac{x}{1-x} = x'$

Since $-\frac{x}{1-x} \in \mathbb{Q} - \{1\}$ whenever $x \in \mathbb{Q} - \{1\}$ thus each element in $\mathbb{Q} - \{1\}$ has its inverse in $\mathbb{Q} - \{1\}$.

(v) For any two elements $x, y \in \mathbb{Q} - \{1\}$,
$$x * y = x + y - xy \text{ and } y * x = y + x - yx = x + y - xy$$
thus $x * y = y * x$, i.e. the elements in $\mathbb{Q} - \{1\}$ satisfy commutative law for the given operation $*$.

Thus the pair $(\mathbb{Q} - \{1\}, *)$ under $*$ as given is an abelian group whose order is infinite as $\mathbb{Q} - \{1\}$ is an infinite set.

Proceed exactly the same way for parts (b), (c) and (d).

For part (b), $e = 0$ and inv of x, $x' = -\dfrac{x}{1+x}$.

For part (c), $e = -1$ and $x' = -x - 2$

For part (d), $e = 2$ and $x' = \dfrac{4}{x}$.

Problem 1.5 : Assuming that the following pairs form groups under the binary operations as given, find identity element e and the inverse for element x in each case :

(a) $(\mathbb{R} - \left\{\dfrac{1}{3}\right\}, *)$, where $*$ is defined as
$$a * b = a + b - 3ab, \text{ for all}$$
$$a, b \in \mathbb{R} - \left\{\dfrac{1}{3}\right\}.$$

(b) $(\mathbb{R} - \{-2\}, *)$ where $*$ is defined as
$$a * b = 2(a + b + 1) + ab, \text{ for all}$$
$$a, b \in \mathbb{R} - \{-2\}.$$

(c) $(\mathbb{R} - \{2\}, *)$ where $*$ is defined as
$$a * b = ab - 2a - 2b + 6, \text{ for all}$$
$$a, b \in \mathbb{R} - \{2\}.$$

Solution : (a) (i) For any $x \in \mathbb{R} - \left\{\dfrac{1}{3}\right\}$, we have by definitions of e_L and given $*$,
$$e_L * x = x$$
$$\Rightarrow e_L + x - 3e_L x = x$$
$$\Rightarrow e_2(1 - 3x) = 0 \Rightarrow e_L = 0 \quad \left(x \neq \dfrac{1}{3}\right)$$

Similarly $x * e_R = x$
$$\Rightarrow x + e_R - 3x e_R = x$$
$$\Rightarrow e_R(1 - 3x) = 0$$
$$\Rightarrow e_R = 0$$

Thus $e_L = e_R = 0 = e$

Binary Operations and Groups 37

(ii) For any $x \in \mathbb{R} - \left\{\frac{1}{3}\right\}$ we have by definitions of x'_L and $*$,

$$x'_L * x = e_L$$
$$\Rightarrow x'_L + x - 3x'_L x = 0$$
$$\Rightarrow x'_L (1 - 3x) = -x$$
$$\Rightarrow x'_L = -\frac{x}{1-3x} \quad \left(x \neq \frac{1}{3}\right)$$

Similarly
$$x * x'_R = e_R$$
$$\Rightarrow x + x'_R - 3xx'_R = 0$$
$$\Rightarrow x'_R (1 - 3x) = -x$$
$$\Rightarrow x'_R = -\frac{x}{1-3x}$$

Hence $x'_L = x'_R = -\dfrac{x}{1-3x} = x'$, the inv of x.

For example if $x = 1/4$, $x' = -1$.

Parts (b) and (c) are left as an exercise for the readers.

Problem 1.6 : Show that the set $G = \{a + \sqrt{2}b : a, b \in \mathbb{R}\}$ forms an infinite abelian group under $* = +$.

Solution : (i) Let $x, y \in G$ so that

$$x = r_1 + \sqrt{2}\, s_1 \text{ and}$$
$$y = r_2 + \sqrt{2}\, s_2, \quad \text{where} \quad r_1, r_2, s_1, s_2 \in \mathbb{R}$$

Then

$$x + y = (r_1 + r_2) + \sqrt{2}\,(s_1 + s_2)$$ also belongs to G, as $r_1 + r_2$ and $s_1 + s_2$ belong to R. Thus + is a b.o. over given set G.

(ii) As elements of G are real numbers, they satisfy AL.

(iii) For any element $x \in G$, since

$$(0 + \sqrt{2}\, 0) + x = x \Rightarrow e_L = 0 + \sqrt{2}\, 0 = 0$$

and $x + (0 + \sqrt{2}\ 0) = x \Rightarrow e_R = 0 + \sqrt{2}\ 0 = 0$
so that $e_L = e_R = 0 = e$ which belongs to G.

(iv) For $x = r_1 + \sqrt{2}\ s_1$, we have
$(-r_1) + \sqrt{2}\ (-s_1)$ satisfying
$$\{(-r_1) + \sqrt{2}\ (-s_1)\} + (r_1 + \sqrt{2}\ s_1) = 0$$
and $(r_1 + \sqrt{2}\ s_1) + \{(-r_1) + \sqrt{2}\ (-s_1)\} = 0$
where $r_1, s_1 \in \mathbb{R}$, thus
$$x'_L = x'_R = (-r_1) + \sqrt{2}\ (-s_1) = x', \text{ and } x' \in G,$$
where $x = r_1 + \sqrt{2}\ s_1$,

Each element in G has its inverse in G.

(v) For any $x, y \in G$, since
$$x + y = (r_1 + \sqrt{2}\ s_1) + (r_2 + \sqrt{2}\ s_2)$$
$$= (r_1 + r_2) + \sqrt{2}\ (s_1 + s_2)$$
$$= (r_2 + r_1) + \sqrt{2}\ (s_2 + s_1)$$
$$= (r_2 + \sqrt{2}\ s_2) + (r_1 + \sqrt{2}\ s_1)$$
$$= y + x$$

$\Rightarrow$ elements of G satisfy commutative law. Since given G is an infinite set, (G +) is an infinite abelian group.

Problem 1.7 : Show that the set $G = \{\ldots m^{-3}, m^{-2}, m^{-1}, m^0, m^1, m^2, m^3, \ldots\}$, where m is a positive integer, forms an infinite abelian group under $* = \cdot$

Solution : (i) Let $x, y \in G$ so that
$$x = m^r \quad \text{and} \quad y = m^s$$
where $r, s \in \mathbb{Z}$ then $x \cdot y = m^r \cdot m^s = m^{(r+s)}$ also belongs to G, as $r + s \in \mathbb{Z}$. Thus $* = \cdot$ is a b.o. over given set G.

(ii) As elements of G are rational numbers, they satisfy AL under $* = \cdot$

(iii) For any element $x \in G$, since $1 \cdot x = x \Rightarrow e_L = 1$, and $x \cdot 1 = x \Rightarrow e_R = 1$ so that $e_L = e_R = 1 = e$. Since $1 = m^0$ ($* = \cdot$) thus $e \in G$.

(iv) For any element $x \in G$, since $x = m^r$, we have m^{-r} satisfying

$$m^{-r} \cdot m^r = e_L = 1 \Rightarrow x'_L = m^{-r}$$

and

$$m^r \cdot m^{-r} = e_R = 1 \Rightarrow x'_R = m^{-r}$$

thus

$$x'_L = x'_R = m^{-r} = x', \text{ the inv of } x,$$

where $x = m^r$. Since $m^{-r} \in G$,

each element in G has its inverse in G.

(v) For any $x, y \in G$, since

$$x \cdot y = m^r \cdot m^s = m^{r+s} = m^{s+r} \quad (r, s \in \mathbb{Z})$$
$$= m^s \cdot m^r$$
$$= y \cdot x$$

$\Rightarrow$ elements of G satisfy commutative law. Since given G is an infinite set, $(G, \cdot)$ is an infinite abelian group.

Note : This problem illustrates a general situation of such groups discussed in chapter III.

Problem 1.8 : Show that the set of 2×2 matrices

$$G = \left\{ \begin{bmatrix} \cos \theta & -\sin \theta \\ \sin \theta & \cos \theta \end{bmatrix} : \theta \in \mathbb{R} \right\}$$

forms an infinite group under matrix multiplication as the operation. Is this group abelian ?

Solution : Let A_α and A_β be any two elements of G, where

$$A_\alpha = \begin{bmatrix} \cos \alpha & -\sin \alpha \\ \sin \alpha & \cos \alpha \end{bmatrix}, \text{ and}$$

$$A_\beta = \begin{bmatrix} \cos \beta & -\sin \beta \\ \sin \beta & \cos \beta \end{bmatrix}; \alpha, \beta \in \mathbb{R}$$

$*$ = Matrix multiplication.

(i) $$A_\alpha A_\beta = \begin{bmatrix} \cos(\alpha+\beta) & -\sin(\alpha+\beta) \\ \sin(\alpha+\beta) & \cos(\alpha+\beta) \end{bmatrix}$$

Since $\alpha + \beta \in \mathbb{R}$, thus $A_\alpha A_\beta \in G$ so that operation is closed.

(ii) Associativity follows from the properties of 2×2 square matrices.

(iii) We have A_0 in G, where

$$A_0 = \begin{bmatrix} \cos 0 & -\sin 0 \\ \sin 0 & \cos 0 \end{bmatrix} = \begin{bmatrix} 1 & 0 \\ 0 & 1 \end{bmatrix}$$

such that

$$A_0 A_\alpha = \begin{bmatrix} 1 & 0 \\ 0 & 1 \end{bmatrix} \begin{bmatrix} \cos\alpha & -\sin\alpha \\ \sin\alpha & \cos\alpha \end{bmatrix} = \begin{bmatrix} \cos\alpha & -\sin\alpha \\ \sin\alpha & \cos\alpha \end{bmatrix} = A_\alpha$$

and similarly

$A_\alpha A_0 = A_\alpha$, so that $e_L = e_R = A_0 = e$, which is present in G.

(iv) For any $\alpha \in \mathbb{R}$,

$$|A_\alpha| = \begin{vmatrix} \cos\alpha & -\sin\alpha \\ \sin\alpha & \cos\alpha \end{vmatrix} = 1,$$

which is non-zero. Thus each element of G is a non-singular matrix and hence will have inverse. The inverse of A_α will be given by $A_\alpha^{-1} = \dfrac{\text{Adj } A_\alpha}{|A|} = \text{Adj } A_\alpha$

$$= \begin{bmatrix} \cos\alpha & \sin\alpha \\ -\sin\alpha & \cos\alpha \end{bmatrix} = \begin{bmatrix} \cos(-\alpha) & -\sin(-\alpha) \\ \sin(-\alpha) & \cos(-\alpha) \end{bmatrix}$$

so that $A_\alpha^{-1} \in G$, as $-\alpha \in \mathbb{R}$,
and

$$A_\alpha^{-1} A_\alpha = \begin{bmatrix} \cos(-\alpha) & -\sin(-\alpha) \\ \sin(-\alpha) & \cos(-\alpha) \end{bmatrix} \begin{bmatrix} \cos\alpha & -\sin\alpha \\ \sin\alpha & \cos\alpha \end{bmatrix}$$

Binary Operations and Groups

$$= \begin{bmatrix} 1 & 0 \\ 0 & 1 \end{bmatrix} = A_0 = e,$$

$$A_\alpha A_\alpha^{-1} = \begin{bmatrix} \cos\alpha & -\sin\alpha \\ \sin\alpha & \cos\alpha \end{bmatrix} \begin{bmatrix} \cos(-\alpha) & -\sin(-\alpha) \\ \sin(-\alpha) & \cos(-\alpha) \end{bmatrix}$$

$$= \begin{bmatrix} 1 & 0 \\ 0 & 1 \end{bmatrix} = A_0 = e,$$

showing that $A'_\alpha = \begin{bmatrix} \cos(-\alpha) & -\sin(-\alpha) \\ \sin(-\alpha) & \cos(-\alpha) \end{bmatrix} \in G$

Hence each element in G has its inverse in G.

(v) Finally

$$A_\alpha A_\beta = \begin{bmatrix} \cos\alpha & -\sin\alpha \\ \sin\alpha & \cos\alpha \end{bmatrix} \begin{bmatrix} \cos\beta & -\sin\beta \\ \sin\beta & \cos\beta \end{bmatrix}$$

$$= \begin{bmatrix} \cos(\alpha+\beta) & -\sin(\alpha+\beta) \\ \sin(\alpha+\beta) & \cos(\alpha+\beta) \end{bmatrix}$$

and similarly

$$A_\beta A_\alpha = \begin{bmatrix} \cos(\alpha+\beta) & -\sin(\alpha+\beta) \\ \sin(\alpha+\beta) & \cos(\alpha+\beta) \end{bmatrix}$$

so that $A_\alpha A_\beta = A_\beta A_\alpha$, *i.e.* elements in G obey commutative law. Since $\alpha \in \mathbb{R}$, set G is infinite. Thus (G, *) is an infinite abelian group.

Problem 1.9 : Show that the set of 2×2 matrices

$$G = \left\{ \begin{bmatrix} a & -b \\ b & a \end{bmatrix} : a, b \in \mathbb{R} \text{ and } a^2 + b^2 \neq 0 \right\}$$

forms an infinite abelian group under matrix multiplication as the operation.

Solution : Proceed as in Problem 1.8. (Infact Problem 1.8 is a special form of this problem with $a = \cos\alpha$, $b = \sin\alpha$).

Problem 1.10 : Show that the set of 2×2 matrices.

$G = \{I_1, I_2, I_3, I_4\}$, where

$$I_1 = \begin{bmatrix} 1 & 0 \\ 0 & 1 \end{bmatrix}, I_2 = \begin{bmatrix} -1 & 0 \\ 0 & 1 \end{bmatrix}, I_3 = \begin{bmatrix} 1 & 0 \\ 0 & -1 \end{bmatrix} \text{ and}$$

$$I_4 = \begin{bmatrix} -1 & 0 \\ 0 & -1 \end{bmatrix}$$

forms a finite abelian group under matrix multiplication as the operation.

Solution : (Hint) All groups axioms for an abelian group can be checked from the following group table containing elements of G :

*Matrix multi.	I_1	I_2	I_3	I_4
I_1	I_1	I_2	I_3	I_4
I_2	I_2	I_1	I_4	I_3
I_3	I_3	I_4	I_1	I_2
I_4	I_4	I_3	I_2	I_1

Problem 1.11 : Give an example to show that the set $v_{n \times n}(\mathbb{Z})$ of all $n \times n$ square non-singular matrices with their elements as integers does not form a group under matrix multiplication.

Solution : From § 1.4.1, example-2, Note-1, on 'Groups of matrices',

$$A = \begin{bmatrix} 2 & 1 \\ 3 & 4 \end{bmatrix} \in v_{2 \times 2}(\mathbb{Z})$$

as A is a 2×2 square matrix with its elements as integers and $|A| = 5 \neq 0$, but inv of $A = A' = \dfrac{\text{Adj } A}{|A|}$

$$= \begin{bmatrix} 4/5 & -1/5 \\ -3/5 & 2/5 \end{bmatrix} \notin v_{2 \times 2}(\mathbb{Z})$$

as its elements are rational numbers. Thus elements of $v_{2 \times 2}(\mathbb{Z})$ may not have their inverses in $v_{2 \times 2}(\mathbb{Z})$, and hence $v_{2 \times 2}(\mathbb{Z})$ can not form a group. This is true for any set $v_{n \times n}(\mathbb{Z})$ of $n \times n$ non-singular matrices with their elements as integers.

Problem 1.12 : Show that the set of nth roots of unity forms a finite abelian group of order n under multiplication.

Solution : Let G be the set of nth roots of unity.
Since $1^{1/n} = (\cos 0 + i \sin 0)^{1/n} = (\cos 2r\pi + i \sin 2r\pi)^{1/n}$

$$= \cos\left(\frac{2r\pi}{n}\right) + i \sin\left(\frac{2r\pi}{n}\right),$$

$$= e^{i(2r\pi/n)} \quad r = 0, 1, 2, \ldots$$

Taking $r = 0, 1, 2, \ldots n - 1$, we get set G as

$$G = \{1, e^{i(2\pi/n)}, e^{i(4\pi/n)}, e^{i(6\pi/n)}, \ldots, e^{i(2(n-1)\pi/n)}\}$$

To verify that G satisfies group axioms of an abelian group wrt * taken as multiplication

(i) Take any two elements $x, y \in G$ so that

$$x = e^{i(2l\pi/n)} \quad \text{and} \quad y = e^{i(2m\pi/n)}$$

where l and m satisfy $0 \le l \le n - 1$ and $0 \le m \le n - 1$ and the maximum value of $l + m$ will be $(n - 1) + (n - 1)$ i.e. $2n - 2$ when $l = n - 1$ and $m = n - 1$. Then

$$xy = e^{i2(l+m)\pi/n}$$

Now two cases arise :

If $l + m \le n - 1$ then $xy \in G$, and
If $l + m > n - 1$ then assume that

$$l + m = n + k \quad \text{where} \quad 0 \le k \le n - 2$$

So that minimum value of $l + m$ is n which is at $k = 0$ and maximum value of $l + m$ is $2n - 2$ which is at $k = n - 2$.

Thus $\quad xy = e^{i2(n+k)\pi/n} = e^{i2\pi} \cdot e^{i2k\pi/n} = e^{i2k\pi/n}$

or $\quad xy = e^{i2k\pi/n}, \quad \text{where} \quad 0 \le k \le n - 2$

hence $\quad xy \in G$, i.e. operation is closed.

(ii) Associativity of elements of G, which are complex numbers, follows from the properties of set $\mathbb{C}$ of all complex numbers.

(iii) Since $\quad e^{i 2\pi \cdot 0/n} = e^0 = 1 \in G$

and $\quad 1 \cdot x = x \cdot 1 = x$

for any x G thus identity element is 1.

(iv) If $x = e^{i\, 2l\pi/n} \in G$ then
$x' = e^{i\, 2(n-l)\pi/n}$ also belongs to G and since
$$x'x = e^{i\, 2\pi/n\, (l+n-l)} = e^{i2\pi} = 1,$$
$$xx' = e^{i\, 2l\pi/n}\, e^{i\, 2(n-l)\pi/n} = e^{i2\pi} = 1$$
thus $x' = e^{i\, 2(n-l)\pi/n}$ is the inverse of $x = e^{i\, 2l\pi/n}$. Thus each element of G has its inverse in G.

Finally, if $x, y \in G$ where
$$x = e^{i\, 2l\pi/n} \text{ and } y = e^{i\, 2m\pi/n}$$
we have
$$xy = e^{i\, 2\pi/n\, (l+m)}$$
$$= e^{i\, 2\pi/n\, (m+l)}$$
$$= e^{i\, 2m\pi/n}\, e^{i\, 2l\pi/n}$$
$$= yx$$
so that elements in G obey commutative law. Since G has n elements, thus G forms a finite abelian group of order n.

Note : The sets
$$A = \{1,\, e^{i\, 2\pi/3},\, e^{i\, 4\pi/3}\} \text{ and}$$
$$B = \{1,\, e^{i\, 2\pi/4},\, e^{i\, 4\pi/4},\, e^{i\, 6\pi/4}\}$$
obtained by putting $n = 3$ and $n = 4$ respectively in G are the sets of cube roots of unity and fourth roots of unity usually written as
$$A = \{1,\, \omega,\, \omega^2\} \text{ and } B = \{1,\, i,\, -i,\, -1\}$$

Thus A and B are finite abelian groups of orders 3 and 4 respectively.

Problem 1.13 : Show that the set $G = \{1, i, -1, -i, j, k, -j, -k\}$ of eight elements with operation $*$ defined over G as $i * i = j * j = k * k = -1$, $i * j = k = -j \times i$, $j \times k = i = -k * j$ and $k * i = j = -i * k$, is a non-abelian group.

Solution : We will check the group axioms from the following composition table made according to given definition of $*$:

*	1	i	−1	−i	j	k	−j	−k
1	1	i	−1	−i	j	k	−j	−k
i	i	−1	−i	1	k	−j	−k	j
−1	−1	−i	1	i	−j	−k	j	k
−i	−i	1	i	−1	−k	+j	k	−j
j	j	−k	−j	k	−1	i	1	−i
k	k	j	−k	−j	−i	−1	i	1
−j	−j	k	j	−k	+1	−i	−1	i
−k	−k	−j	k	j	i	1	−i	−1

(i) It is evident from the table that the given operation is closed.

(ii) Here $e = 1$, as $1 * x = x * 1 = x$ for any x from the given set.

(iii) For any three elements of the set, associativity can be checked. For example $i * (j * k) = (i * j) * k$, as from the table

$$i * (j * k) = i * (i) = -1, \text{ and}$$
$$(i * j) * k = k * k = -1.$$

(iv) Inv 1 = 1, Inv $i = -i$, Inv $-1 = -1$, Inv $-i = i$, Inv $j = -j$, Inv $k = -k$, Inv $-j = j$ and Inv $-k = k$.

Thus each element of the set has its inverse in the set.

(v) Finally for any $x, y \in G$, $x * y \neq y * x$. For example from the table $-i * k = j$ and $k * (-i) = -j$.

Thus (G, *) is a finite non-abelian group of order 8. The group is called the *Hamiltonian group*.

Problem 1.14 : Show that the set $G = \{(x, y) : x, y \in \mathbb{R} \text{ and } x \neq 0\}$ of all ordered pairs (x, y) of real numbers for which $x \neq 0$ forms on infinite group wrt operation * defined by

$$(a, b) * (c, d) = (ac, bc + d), \forall a, b, c, d \in \mathbb{R} \text{ and } a \neq 0, c \neq 0.$$

Solution : (i) Since $a \neq 0$, $c \neq 0$, therefore $ac \neq 0$ and ac, $bc + d \in \mathbb{R}$ as $a, b, c, d \in \mathbb{R}$. Hence $(ac, bc + d) \in G$ and the given operation is a b.o. over G.

(ii) For any 3 elements (a, b), (c, d) and (i, j) of set G, where $a \neq 0$, $c \neq 0$ and $i \neq 0$, we have

$$[(a, b) * (c, d)] * (i, j) = (ac, bc + d) * (i, j)$$
$$= (aci, (bc + d) i + j)$$
$$= (aci, bci + di + j), \text{ and}$$
$$(a, b) * [(c, d) * (i, j)] = (a, b) * [(ci, di + j)]$$
$$= (aci, bci + di + j),$$

so that

$$[(a, b) * (c, d)] * (i, j) = (a, b) * [(c, d) * (i, j)]$$

or, the given operation $*$ obeys AL.

(iii) We have the element $(1, 0)$ in the set G and for any element $(a, b) \in G$, since

$$(1, 0) * (a, b) = (a, b) \text{ as well as}$$
$$(a, b) * (1, 0) = (a, b)$$

showing that $(1, 0)$ is the identity element for the operation $*$.

(iv) For any $(a, b) \in G$, let (a', b') be the inverse of, (a, b). Then by definition of inverse

$$(a', b') * (a, b) = (1, 0)$$
$\Rightarrow \qquad (a' a, b'a + b) = (1, 0)$
$\Rightarrow \qquad a'a = 1 \text{ and } b'a + b = 0$
$\Rightarrow \qquad a' = 1/a \text{ and } b' = -b/a$

Similarly

$$(a, b) * (a', b') = (1, 0) \Rightarrow a' = 1/a, b' = -b/a$$

Thus $(1/a, -b/a)$ is inverse for (a, b) and since $1/a \neq 0$ as $a \neq 0$, $(-1/a, -b/a) \in G$. Thus each element of G has its inverse in G.

By (1) – (iv), (G, $*$) is a group.

Further, G has infinite elements and since $(c, d) * (a, b) = (c, da + b) \neq (a, b) * (c, d)$, $(G, *)$ is infinite non-abelian group.

Problem 1.15 : Show that the set $G = \{(x, y) : x, y \in \mathbb{R}$ and $y \neq 0\}$ of all ordered pairs (x, y) of real numbers for which $y \neq 0$ forms an infinite group wrt operation $*$ defined by $(a, b) * (c, d) = (a + bc, bd)$ for all $a, b, c, d \in \mathbb{R}$ and $b, d \neq 0$. Is this group abelian ?

Solution : Proceed exactly as in above problem.

Problem 1.16 : Let $G = \{f_i = 1, 2, \ldots 6\}$ be the set of six functions over $\mathbb{C}$ defined by

$$f_1(z) = z, f_2(z) = 1/z, f_3(z) = 1 - z, f_4(z) = z/z - 1$$
$$f_5(z) = 1/1 - z \text{ and } f_6(z) = z - 1/z, z \in \mathbb{C}.$$

Over G, define the operation $*$ as the composite of two functions (*i.e.* substitution of one function into the other). Show that $(G, *)$ is a finite non-abelian group of order 6.

Solution : Consider $(f_4 * f_3)(z)$

$$= f_4(f_3(z)) \quad \text{(By definition of *)}$$

$$= \frac{f_3(z)}{f_3(z) - 1} \quad \text{(By definition of } f_4(z))$$

$$= \frac{(1-z)}{(1-z) - 1} = \frac{z-1}{z} = f_6(z),$$

or $\quad f_4 * f_3 = f_6 \in G$

Computing all such products $f_i * f_j$ for any $f_i, f_j \in G$, we see that $f_i * f_j \in G$, *i.e.* $*$ is a b.o. over G and we have the following b.o. table :

$*$	f_1	f_2	f_3	f_4	f_5	f_6
f_1	f_1	f_2	f_3	f_4	f_5	f_6
f_2	f_2	f_1	f_5	f_6	f_3	f_4
f_3	f_3	f_6	f_1	f_5	f_4	f_2
f_4	f_4	f_5	f_6	f_1	f_2	f_3
f_5	f_5	f_4	f_2	f_3	f_6	f_1
f_6	f_6	f_3	f_4	f_2	f_1	f_5

From above table we notice the following :

(i) Elements f_i ($i = 1, 2, ..., 6$) satisfy A.L. under $*$. For example

$$f_2 * (f_3 * f_4)(z) = f_2 * (f_3(f_4(z)))$$
$$= f_2 \{f_3(f_4(z))\}$$
$$= \frac{1}{f_3(f_4(z))}$$
$$= \frac{1}{1 - f_4(z)}$$
$$= \frac{1}{1 - \frac{z}{z-1}} = -(z - 1),$$

and $((f_2 * f_3) * f_4)(z) = (f_2 * f_3)(f_4(z))$
$$= f_2 * (f_3(f_4(z)))$$
$$= f_2 \{f_3(f_4(z))\}$$
$$= -(z - 1), \text{ showing that}$$
$$f_2 * (f_3 * f_4) = (f_2 * f_3) * f_4$$

(iii) From above table, for any $f_i \in G$,
$$f_i * f_i = f_i * f_i = f_i.$$
so that f_1 is identity element. Also, from direct calculations,
$$(f_1 * f_5)(z) = f_1 * (f_5(z)) = f_1 \{f_5(z)\} = f_5(z),$$
and $(f_5 * f_1)(z) = f_5 * f_1(z) = f_5\{f_1(z)\} = \frac{1}{1 - f_1(z)}$
$$= \frac{1}{1 - z} = f_5(z), \text{ thus}$$
$$f_1 * f_5 = f_5 * f_1 = f_5.$$

(iii) From the table,
$$\text{Inv } f_1 = f_1, \text{ Inv } f_2 = f_2, \text{ Inv } f_3 = f_3,$$
$$\text{Inv } f_4 = f_4, \text{ Inv } f_5 = f_6 \text{ and Inv } f_6 = f_5.$$

Also, from direct calculations,

$$(f_6 * f_5)(z) = f_6 * (f_5(z)) = f_6(f_5(z)) = \frac{f_5(z) - 1}{f_5(z)}$$

$$= \frac{\frac{1}{1-z} - 1}{\frac{1}{1-z}} = z = f_1(z)$$

or, $\qquad f_6 * f_5 = f_1$

Similarly $\quad f_5 * f_6 = f_1$

so that $\qquad$ inv $f_5 = f_6$ and inv $f_6 = f_5$.

(iv) Finally from the table or by direct calculations,

$f_i * f_j \neq f_j * f_i$ for any $f_i, f_j \in G$.

Thus $(G, *)$ is a finite non-abelian group of order 6.

Note : Here a natural question arises whether similar sets of order less than 6 be constructed to form a non-abelian group of order less than 6. The answer is no. Consider following two problems.

Problem 1.17 : Let $G = \{f_i : i = 1, 2, 3, 4\}$ be the set of four functions over $\mathbb{C}$ defined by $f_1(z) = z$, $f_2(z) = -z$, $f_3(z) = \frac{1}{z}$ and $f_4(z) = -\frac{1}{z}$, $z \in \mathbb{C}$. Define operation $*$ over G as in problem 1.16. Show that $(G, *)$ forms a finite abelian group of order 4.

Problem 1.18 : Let $G = \{f_i : i = 1, 2, ..., 6\}$ be the set of six functions over $\mathbb{C}$ defined by $f_1(z) = z$, $f_2(z) = \frac{1}{1-z}$, $f_3(z) = \frac{z-1}{z}$, $f_4(z) = \frac{1}{z}$, $f_5(z) = 1-z$ and $f_6(z) = \frac{z}{z-1}$, $z \in \mathbb{C}$. Define operation $*$ over G as in problem 1.16. Show that $(G, *)$ forms a finite non-abelian group of order 6.

Solution : For above two problems compute the b.o. tables and proceed as in problem 1.16.

Problem 1.19 : Show that the set B_A of all one-to-one onto mappings from a non-empty set A to itself forms a group under composition of maps as the operation. Show also that this group may not be abelian.

Solution : Follow the steps given in C (Groups of functions) of Illustrative examples 1.4.1. The group B_A is the *transformation group* of A.

Problem 1.20 : Show that the set $G = \{0, 1, 2, ..., m - 1\}$ of first m non-negative integers is a group under addition modulo m as the operation. Show also that the group is finite abelian group.

Solution : Recall the definition of addition modulo a positive integer m. For any integers a and b, we write $a +_m b = r$ if r is the remainder when the ordinary sum $a + b$ is divided by m. Thus $0 \leq r < m$. [See Prelim. 2 (v)]

We will check all the group axioms for an abelian group for G under $* = +_m$.

(1) If $a, b \in G$ then $a +_m b = r \in G$, as $0 \leq r < m$.

(2) Associativity of elements of G follow from the properties of $+_m$ [See Prelim. 2 (v)].

(3) Since $0 +_m x = x +_m 0 = x$, for any $x \in G$, identity element is 0 which is present in G.

(4) For any non-zero element $y \in G$, since

$(m - y) +_m y = y +_m (m - y) = 0$,

so that inv of y, $y' = m - y$. Also inv of $0 = 0$. Thus each element of G has its inverse in G.

(5) In general, for any two integers a and b, $a +_m b = b +_m a$, thus $+_m$ satisfies commutative law.

Thus G, which has m elements, forms a finite abelian group under $* = +_m$.

Problem 1.21 : Show that the set $G = \{1, 2, 3, ..., p - 1\}$ of $(p - 1)$ integers where p is a prime number is a group under multiplication modulo p as the operation. Show also that the group is finite abelian group.

Solution : Recall the definition of multiplication modul p for any integers a and b. We write $a \times_p b = r$ if r is th

Binary Operations and Groups 51

remainder when the ordinary product ab is divided by p. Thus $0 \leq r < p$ {See Prelim. 2 (v)]

We will check all the group axioms for an abelian group for G under $* = \times_p$.

(1) If $a, b \in G$ then $a \times_p b = r \in G$, as $1 \leq r < p$.

(2) Associativity of element of G follows from the properties of $\times_p$ [See Prelim. 2 (v)].

(3) Since $1 \times_p x = x \times_p 1 = x$, for any $x \in G$, identity element is 1 and is present in G.

(4) Since p is prime, for any $x \in G$ we will always get some $x' \in G$ satisfying

$x' \times_p x = x \times_p x' = 1$,

i.e. inv of $x = x'$ for any $x \in G$. Thus each element of G has its inverse in G.

(5) In general, for any two integers a and b, $a \times_p b = b \times_p a$, thus $\times_p$ satisfies commutative law.

Thus G, which has $(p - 1)$ elements, forms a finite abelian group under $* = \times_p$.

Note : 1. If in the above result, p is a composite integer (in place of a prime number) then G will not form a group under $\times_p$ as for any $a, b \in G$, $a \times_p b = 0 \notin G$. Even if we include 0 in the set G, G will not be able to form a group as, then, inv. of 0 wrt $\times_p$ does not exist.

2. Apart from forming groups with sets $\{0, 1, 2, ..., p - 1\}$ when p is a prime number, the operation of multiplication modulo a composite integer can also form groups with some finite sets as we have shown in the problem 1.2.4.

Problem 1.22 : Show that the set G = $\{0, 1, 2, 3\}$ of first four non-negative integers forms a group under $* = +_4$.

Solution : (This problem illustrates the problem 1.20 when $m = 4$)

For the operation $* = +_4$ over given set G, we have the following composition table which is easily seen to be a b.o. table :

* = +₄	0	1	2	3
0	0	1	2	3
1	1	2	3	0
2	2	3	0	1
3	3	0	1	2

Here $e = 0$. Inv of $1 = 3$, Inv of $2 = 2$, and Inv of $3 = 1$.

All group axioms can be checked from this table. Notice that elements in the above table move in a cyclic order and this group is the group $\mathbb{Z}_4$ (See 1.5.1 (3)), and it is abelian. (Its table is symmetrical wrt principal diagonal).

Problem 1.23: Show that the set $G = \{1, 2, 3, 4\}$ of first four positive integers forms a group under $* = \times_5$.

Solution: (This problem illustrates problem 1.21 when $p = 5$)

Since $* = \times_5$ is a b.o. over $\mathbb{Z}$, computing the b.o. table over given G we have:

$* = \times_5$	1	2	3	4
1	1	2	3	4
2	2	4	1	3
3	3	1	4	2
4	4	3	2	1

Here $e = 1$, Inv $1 = 1$, Inv $2 = 3$, Inv $3 = 2$ and Inv $4 = 4$ Associative law can easily be checked, hence the table is a group table. Also, it is symmetrical about principal diagonal hence group $(G, \times_5)$ is a finite abelian group.

Problem 1.24: Show that following pairs form finite abelian groups:

(i) $G = \{2, 4, 6, 8\}$, $* = \times_{10}$

(ii) $G = \{3, 6, 9, 12\}$, $* = \times_{15}$, and

(iii) $G = \{2, 4, 6, 8, 10, 12\}$, $* = \times_{14}$.

Binary Operations and Groups

Solution : (*i*) By actually performing the operation $* = \times_{10}$ over the given set $G = \{2, 4, 6, 8\}$, we get following composition table :

$* = \times_{10}$	2	4	6	8
2	4	8	2	6
4	8	6	4	2
6	2	4	6	8
8	6	2	8	4

The above is a b.o. table as for any $x, y \in G$, $x \times_{10} y \in G$.

Here $e = 6$ (Notice that the elements below 6 in the third column and those against 6 in the third row appear in the same sequence as they appear in the set G, *i.e.*, in the sequence 2, 4, 6, 8.) Inv of 2 = 8, Inv 4 = 4, Inv of 6 = 6 and Inv of 8 = 2. For any three elements $x, y, z \in G$, we have

$$x \times_{10} (y \times_{10} z) = (x \times_{10} y) \times_{10} z$$

For example, from the table

$$4 \times_{10} (8 \times_{10} 2) = 4 \times_{10} (6) = 4, \text{ and}$$

$$(4 \times_{10} 8) \times_{10} 2 = 2 \times_{10} 2 = 4.$$

Thus the above table is a group table and hence $(G, \times_{10})$ **is a finite group of order 4. Further, the table is symmetrical wrt principal diagonal, this group is abelian as well. Follow above and do parts (ii) and (iii) yourself.**

Problem 1.25 : Show that the set of residue classes modulo a positive integer m is an abelian group of order m under addition of residue classes as the operation.

Solution : Recall the definition of residue classes modulo m [See Prelim. 2 (*v*)]

If $n \in \mathbb{Z}$ then the residue class $\bar{n}$ modulo m of the set $\mathbb{Z}$ is given by

$$\bar{n} = \{x : x \in \mathbb{Z} \text{ and that } x - n \text{ is divisible by } m\}$$

The set of all such residue classes modulo m of the set $\mathbb{Z}$, denoted by $\mathbb{Z}_m$, will have m distinct elements and they are the residue classes $\bar{0}, \bar{1}, \bar{2}, ..., \overline{m-1}$ modulo m of $\mathbb{Z}$. Thus

$$\mathbb{Z}_m = \{\bar{0}, \bar{1}, \bar{2}, \ldots, \overline{m-1}\}$$

The addition of two residue classes $\bar{a}$ and $\bar{b}$, where $a, b \in \mathbb{Z}$, is defined as [See Prelim. 2 (v)],

$$\bar{a} + \bar{b} = \overline{(a+b)}$$

We will check all the group axioms for an abelian group for $\mathbb{Z}_m$ under $* =$ sum of residue classes.

(i) If $\bar{a}, \bar{b} \in \mathbb{Z}_m$ then $\bar{a} + \bar{b} = \overline{a+b} \in \mathbb{Z}_m$, as $a + b \in \mathbb{Z}$.

(ii) For any three elements $\bar{a}, \bar{b}$ and $\bar{c} \in \mathbb{Z}_m$, since

$$\begin{aligned}\bar{a} + (\bar{b} + \bar{c}) &= \bar{a} + \overline{(b+c)} \\ &= \overline{(a+(b+c))} \\ &= \overline{((a+b)+c)} \quad \text{(as } a, b, c \in \mathbb{Z}) \\ &= \overline{(a+b)} + \bar{c} \\ &= (\bar{a} + \bar{b}) + \bar{c}\end{aligned}$$

showing that A.L. is satisfied.

(iii) Since $\bar{0} + \bar{a} = \overline{0+a} = \bar{a}$ and $\bar{a} + \bar{0} = \overline{a+0} = \bar{a}$, for any $\bar{a} \in \mathbb{Z}_m$ thus identity element is the residue class $\bar{0}$, which is present in $\mathbb{Z}_m$.

(iv) For any element $\bar{a}$ (other than $\bar{0}$) of the set $\mathbb{Z}_m$, since

$$\overline{(-a)} + \bar{a} = \overline{(-a+a)} = \bar{0},$$

and $\qquad \bar{a} + \overline{(-a)} = \overline{(a-a)} = \bar{0},$

inv of $\bar{a}$ is $\overline{(-a)}$, which is present in $\mathbb{Z}_m$ as $-a \in \mathbb{Z}$.

In particular if $\bar{r} \in \mathbb{Z}_m$ where $0 \leq r < m$ then inverse of $\bar{r}$ will be the residue class $\overline{(m-r)}$.

(v) For any $\bar{a}, \bar{b} \in \mathbb{Z}_m$, since

$$\bar{a} + \bar{b} = \overline{(a+b)} = \overline{(b+a)} = \bar{b} + \bar{a}, \quad \text{(as } a, b \in \mathbb{Z}).$$

Binary Operations and Groups

addition of residue classes satisfies commutative law.

Thus $\mathbb{Z}_m$, which has m elements, forms a finite abelian group under addition of residue classes.

Problem 1.26: Show that the set of non-zero residue classes modulo a positive prime integer p forms an abelian group of order $(p-1)$ under multiplication of residue classes as the operation.

Solution: The set $\mathbb{Z}_p^*$ of non-zero residue classes modulo a positive prime integer p will have residue classes $\overline{1}, \overline{2}, \overline{3}, ..., \overline{(p-1)}$, as its elements. Thus

$$\mathbb{Z}_p^* = \{\overline{1}, \overline{2}, \overline{3}, ..., \overline{(p-1)}\} = \{\overline{n} : n \in \mathbb{Z} \text{ and } n \text{ is not divisible by } p\}$$

The operation of multiplication of residue classes $\overline{a}, \overline{b}$, where $a, b \in \mathbb{Z}$, is defined as [See Prelim. 2 (v)],

$$\overline{a}\,\overline{b} = \overline{ab}$$

We will check all the group axioms for an abelian group for above set $\mathbb{Z}_p^*$ under multiplication of residue classes :

(i) Let $\overline{a}, \overline{b} \in \mathbb{Z}_p^*$, since $\overline{a}\,\overline{b} = \overline{ab}$ and $ab \in \mathbb{Z} \Rightarrow \overline{ab} \in \mathbb{Z}_p$. Now $\overline{ab} \neq \overline{0}$, for if $\overline{ab} = \overline{0}$ by equality of residue classes [see Prelim. 2 (v)], ab is divisible by p which further implies that a is divisible by p or b is divisible by p, neither of which is true as $\overline{a}, \overline{b} \in \mathbb{Z}_p^*$. Thus $\overline{ab} \neq \overline{0}$ or $\overline{ab} \in \mathbb{Z}_p^*$, *i.e.* operation is closed.

(ii) Take any three elements $\overline{a}, \overline{b}, \overline{c}$ from $\mathbb{Z}_p^*$. Since $\overline{a}\,(\overline{b}\overline{c})$

$$= \overline{a}\,\overline{(bc)}$$
$$= \overline{a\,(bc)} = \overline{(ab)\,c} \quad \text{(as } a, b, c \in \mathbb{Z})$$
$$= \overline{(ab)}\,\overline{c}$$
$$= (\overline{a}\overline{b})\,\overline{c}$$

showing that elements of $\mathbb{Z}_p^*$ obey AL.

(iii) We have $\overline{1} \in \mathbb{Z}_p$ and $\overline{1} \neq \overline{0}$ as 1 is not divisible by p, *i.e.*, $\overline{1} \in \mathbb{Z}_p^*$. Also for any $\overline{a} \in \mathbb{Z}_p^*$

$$\overline{1}\bar{a} = \bar{a}\overline{1} = \bar{a}$$

$\overline{1}$ is identity element of $\mathbb{Z}_p^*$.

(iv) Take an element $\bar{a} \in \mathbb{Z}_p^*$, then by (1) $\overline{1}\bar{a}, \overline{2}\bar{a}, ..., \overline{(p-1)}\bar{a}$ all belong to $\mathbb{Z}_p^*$ and are distinct, for, if

$$\bar{i}\bar{a} = \bar{j}\bar{a}$$

for any $i, j \in \mathbb{Z}$ where $1 \leq i \leq p-1$
$1 \leq j \leq p-1$, and $i > j$.

or if $\qquad \overline{ia} = \overline{ja}$

which implies that $(ia - ja)$ is divisible by p (by equality of cosets) or $(i - j)$ is divisible by p which is not true as $0 < i - j < p$. Thus $\bar{i}\bar{a} \neq \bar{j}\bar{a}$ or the elements $\overline{1}\bar{a}, \overline{2}\bar{a}, ..., \overline{(p-1)}\bar{a}$ are distinct. Since $\overline{1} \in \mathbb{Z}^*_p$.

$\Rightarrow$ for $1 \leq k \leq p-1$, $\bar{k}\bar{a} = \overline{1}$ and $\bar{a}\bar{k} = \overline{1}$ implying that inv of $\bar{a} = \bar{k}$ which is present in $\mathbb{Z}^*_p$. Thus each element of $\mathbb{Z}^*_p$ has its inverse in $\mathbb{Z}^*_p$.

(v) Finally, since for any $\bar{a}, \bar{b} \in \mathbb{Z}^*_p$,

$$\bar{a}\bar{b} = \bar{b}\bar{a},$$

elements in $\mathbb{Z}^*_p$ obey commutative law.

Thus $\mathbb{Z}^*_p$ forms a finite abelian group of order $(p-1)$ under multiplication of residue classes.

Note : If in problem 1.2.6, p is a composite integer (in place of a positive prime), then $\mathbb{Z}^*_p$ will not form a group, as then, $p = ij$ where $1 < i < p$ and $1 < j < p$ and $\bar{p} = \overline{\bar{ij}}$.

Now $\qquad 1 < i < p \Rightarrow i$ is not divisible by p

$$\Rightarrow \bar{i} \neq \bar{0}$$

$$\Rightarrow \bar{i} \in \mathbb{Z}^*_p. \text{ Similarly } \bar{j} \in \mathbb{Z}^*_p$$

Thus $\qquad \bar{v} = \overline{\bar{ij}}$

$\Rightarrow \qquad \bar{0} = \bar{i}\bar{j}, \quad \text{(as } \bar{p} = \bar{0}\text{)}$

$\Rightarrow \qquad \bar{i}\bar{j} \notin \mathbb{Z}^*_p \quad \text{or} \quad \bar{i}, \bar{j} \in \mathbb{Z}^*_p \Rightarrow \bar{i}\bar{j} \notin \mathbb{Z}^*_p$

i.e. the operation is not closed in this case.

Problem 1.27 : Let $m \in \mathbb{Z}^+$ and G be a set of all residue classes $\bar{a}$ where $a \in \mathbb{Z}$ and a is relatively prime to m, i.e.

$$G = \{\bar{a} : a \in \mathbb{Z} \text{ and } \gcd(a, m) = 1\}$$

Show that G forms a group under multiplication of residue classes as the operation.

Solution : Proceed exactly as in problem 1.26.

Problem 1.28 : Show that the set of all polynomials $\mathbb{P}$ of degree n with integral coefficients forms an infinite abelian group under ordinary addition as the operation.

Solution : An element p of the set $\mathbb{P}$ will be

$$p(x) = a_0 + a_1 x + a_2 x^2 + \dots + a_n x^n,$$

where $a_0, a_1, \dots, a_n \in \mathbb{Z}$. By changing the coefficients we get all the elements of $\mathbb{P}$ which can be easily shown to satisfy all group axioms for an abelian group.

Problem 1.29 : In a group $(G, *)$, $(ab)^m = a^m b^m$ true for three consecutive integers m and for all elements $a, b \in G$. Show that G is an abelian group.

Solution : If k is an integer then $k, k+1$ and $k+2$ are three consecutive integers. For any $a, b \in G$, $ab \in G$. Consider

$(ab)^{k+2}$ (Using juxtaposition, i.e. writing ab for $a * b$)

$\qquad\qquad = (ab)^{k+1} (ab)$

Thus $\qquad a^{k+2} b^{k+2} = a^{k+1} b^{k+1} (ab)$ (as given)

$\Rightarrow \qquad a^{k+1} a\, b^{k+1} b = a^{k+1} b^{k+1} ab$

$\Rightarrow \qquad\qquad a b^{k+1} = b^{k+1} a$ (by cancellation laws)

$\Rightarrow \qquad\quad a^k (ab^{k+1}) = a^k (b^{k+1} a)$

$\Rightarrow \qquad\quad a^{k+1} b^{k+1} = a^k b^k ba$ (by AL)

$\Rightarrow \qquad\quad (ab)^k (ab) = (ab)^k ba$

$\Rightarrow \qquad\qquad\quad ab = ba,$ i.e. G is an abelian group.

Problem 1.30 : For the group $(G, \times_{10})$, where $G = \{2, 4, 6, 8\}$, verify the following results :

(i) $\qquad (ab)^{-1} = b^{-1}a^{-1}$
(ii) $\qquad (ab)^2 = a^2b^2$, and
(iii) $\qquad (a^{-1})^{(-1)} = a$

for any elements $a, b \in G$.

Solution : Let $a = 4$ and $b = 8$ then from the group table of $(G, \times_{10})$ (see Problem 1.24)
$$ab = a * b = 4 \times_{10} 8 = 2$$
so that $\qquad (ab)^{-1} = \text{Inv of } 2 = 8$

Also $\qquad a^{-1} = \text{Inv of } 4 = 4$ and $b^{-1} = \text{Inv of } 8 = 2$

so that $\qquad b^{-1}a^{-1} = b^{-1} * a^{-1} = 2 \times_{10} 4 = 8,$

or $\qquad (ab)^{-1} = b^{-1}a^{-1}$

$$(ab)^2 = (a * b)^2 = (a * b) * (a * b)$$
$$= (4 \times_{10} 8) \times_{10} (4 \times_{10} 8)$$
$$= 2 \times_{10} 2$$
$$= 4, \text{ and}$$
$$a^2b^2 = a^2 * b^2 = (a * a) * (b * b)$$
$$= (4 * 4) * (8 * 8)$$
$$= 6 * 4$$
$$= 4$$

so that $\qquad (ab)^2 = a^2b^2$

Finally $\qquad (b^{-1})^{-1} = \text{Inv of } b^{-1} = \text{Inv of (Inv of } b)$
$$= \text{Inv of (Inv of 8)}$$
$$= \text{Inv of (2)}$$
$$= 8$$
$$= b,$$

thus $\qquad (b^{-1})^{-1} = b.$

Problem 1.31 : Let $(G, *)$ be a group with identity e. (i) If for any $a, b, c \in G$, $a * x * b^{-1} * c = e$, find x.

(ii) If for any $a, b \in G$, $b^{-1} * a^{-1} * b * a = e$, show that $ab = ba$, i.e. G is abelian.

Solution : *(i)* $a * x * b^{-1} * c = e$
$\Rightarrow (a * x * b^{-1} * c) * c^{-1} = e * c^{-1} = c^{-1}$
$\Rightarrow \qquad a * x * b^{-1} = c^{-1}$ (by AL)
$\Rightarrow \qquad (a * x * b^{-1}) * b = c^{-1} * b$
$\Rightarrow \qquad a * b = c^{-1} * b$
$\Rightarrow \qquad a^{-1} * (a * x) = a^{-1} * (c^{-1} * b) \Rightarrow x = a^{-1} * c^{-1} * b.$
(ii) $\qquad b^{-1} * a^{-1} * b * a = e$
$\Rightarrow (b^{-1} * a^{-1}) * (b * a) = e$ (generalised AL)
$\Rightarrow (a * b)^{-1} * (b * a) = e \quad (b^{-1} * a^{-1} = (a * b^{-1}))$
$\Rightarrow \qquad b * a = a * b \quad (x^{-1} * y = e \Rightarrow y = x)$
or $\qquad ba = ab$, i.e.

G is abelian.

1.7 Alternative Group-Axioms

In 1.4 we have defined group by means of a set of three postulates, called the group-axioms. In the present section we have arranged a number of results giving us alternative definitions of a group. These results, given in the form of following theorems, are shown to be equivalent to the definition 1.4.5 of a group.

Theorem 1.7.1 : The pair (G, $*$) of a non-empty set G and a b.o. defined over G is group if the following axioms hold good in G :

(G_1') Associative law.

(G_2') Existence of left identity e_L in G.

(G_3') Existence of left inverse a'_L for every element a in G.

Proof : In order to prove the equivalence of this definition with that given in 1.4.5, we need to prove the following two :

(i) The left identity e_L acts as right identity as well, i.e. $a * e_L = a$, for all $a \in G$, and

(*ii*) The left inverse a'_L of element a acts as right inverse of a as well, *i.e.*

$$a * a'_L = e_L \text{ for all } a \in G.$$

As proved in corollaries 1.6.1 and 1.6.2, (G'_1) and (G'_2) imply $e_L = e_R$ which is (*i*) above and (G'_1) and (G'_3) imply $a'_L = a'_R$ which is (*ii*). Hence above three axioms are equivalent to those given in definition 1.4.5.

Note : 1. The result contained in theorem 1.7.1 provides us another definition of a group, called *one-sided definition*. Thus in order to check if a non-empty set G under a b.o. * forms a group, we need to check only the axioms (G'_1), (G'_2) and (G'_3) (called one sided or *left sided axioms*), which are :

(G'_1) $a * (b * c) = (a * b) * c$, for all $a, b, c \in G$

(G'_2) there exists e_L in G s.t.

$e_L * a = a$ for all $a \in G$, and

(G'_3) for every $a \in G$, there exists a'_L in G such that $a'_L * a = e_L$.

2. Theorem 1.7.1 can also be stated in terms of right identity element e_R and right inverse element a'_R of element a. Thus we have another set of three axioms (called *right-sided axioms*) which define a group. They are :

(G''_1) $a * (b * c) = (a * b) * c$, for all $a, b, c \in G$.

(G''_2) there exists e_R in G s.t.

$a * e_R = a$ for all $a \in G$,

and (G''_3) for every $a \in G$, there exists a'_R in G such that $a'_R * a = e_R$.

3. Notice that it is not possible to define a group with mixed axioms (G'_1), (G'_2) and (G''_3) or (G'_1), (G'_3) and (G''_2), *i.e.* we can not define a group with the help of AL alongwith the existence of left identity e_L and right inverse a'_R in G or alongwith the existence of right identity e_R and left inverse a'_L in G. Because in these cases corollaries 1.6.1 and 1.6.2 do not hold at all.

Illustrative Examples 1.3.1 (examples 1 (*v*) and (3) illustrates the situation.

Binary Operations and Groups

Theorem 1.7.2 : A non-empty set G together with an associative b.o. $*$ is a group if and only if the linear equations $a * x = b$ and $y * a = b$, for all $a, b \in G$, have solutions in G.

Proof : (i) *First part* : If (G, $*$) is a group then, as shown in Theorem 1.6.4, linear equations $a * x = b$ and $y * a = b$, $a, b \in G$, have unique solutions in G.

(ii) *Second part* : Conversely let G be a non-empty set, $*$ is associative b.o. over G and the equations $a * x = b$, $y * a = b$ for $a, b \in G$, have solution in G.

The equation $y * a = a$ has a solution in G. Let the solution be $y = e$. Then $e * a = a$ ... (1). Also the equation $a * x = b$ has a solution, say x, in G so that $a * x_1 = b$ ... (2)

Now
$$e * b$$
$$= e * (a * x_1) \quad \text{(by 2)}$$
$$= (e * a) * x_1$$
$$= a * x_1 \quad \text{(by 1)}$$
$$= b \quad \text{(by 2)}$$

$\Rightarrow e * b = b$ for any $b \in G$, i.e. e is the left identity of $*$ and it belong to G. Next, the solution of the equation $y * a = e$, say a' belongs to G so that $a' * a = e$ for any $a \in G$. Thus the element a' is the left inverse of element a, and by Theorem 1.7.1 (G, $*$) is a group.

Corollary 1.7.1 : A non-empty set G together with an associative b.o. $*$ is a group if and only if $a * G = G * a = G$ for all $a \in G$, where $a * G = \{a * x : x \in G\}$ and $G * a = \{x * a : x \in G\}$.

Theorem 1.7.3 : A non-empty finite set G together with an associative b.o. $*$ defined over G is a group if both the cancellation laws hold true in G.

Proof : Since G is finite, let
$$G = \{a_1, a_2, ..., a_i, ..., a_j, ..., a_n\}$$

If $a \in G$ then the elements $a * a_1, a * a_2, ..., a * a_n$ are all distinct and belong to G, for, if $a * a_i = a * a_j$ where $i \neq j$ then by left cancellation law $a_i = a_j$ which contradicts that $i \neq j$. Hence the set G can also be written as G = $\{a * a_1, a * a_2, ..., a * a_n\}$ where the order in which the elements appear is not the same as in G = $\{a_1, a_2, ..., a_n\}$. Thus any element $b \in$ G can also be written as $b = a * a_i$ for some $a_i \in$ G, i.e. the equation $b = a * x$, $a, b \in$ G, has a solution in G. Proceeding the same way, and using right cancellation law, we can show that the equation $b = y * a$ has a solution in G. Thus by theorem 1.7.2 (G, *) is a group.

Note : Theorem 1.7.3 is not true when the set G is infinite. For example, * = · (ordinary multiplication) is an associative b.o over $\mathbb{Z}$, the set of integers, which also satisfies both the cancellation laws, but ($\mathbb{Z}$, ·) is not a group (Why ?).

1.8 Order of an Element

In a group, every element has an order which provides important informations about the group.

Definition 1.8.1 : Let (G, *) be a group, $a \in$ G and n be the least positive integer such that $a^n = e$, where a^n, as given in definition 1.1.4 is given by $a * a * ... * a$ (n times) and e is the identity element of G. Then n is said to be the *order of element* a, denoted by 0 (a). Thus if there exists a positive integer l such that $a^l = e$ and $l \neq n$ then $l > n$. Since $e^1 = e$, order of identity element in a group is always 1. Notice that identity element will be the only element with order 1. Remaining all elements of a group will have their orders greater than 1.

Definition 1.8.2 : If no such positive integer n exists so that $a^n = e$ then the element a is said to be of *infinite order* and we alott zero as its order.

Definition 1.8.3 : If every element of a group G has a finite order then G is said to be a *torsion (or periodic) group*; if every non-identity element has an infinite order then G is said to be a *torsion-free group*. If except for some non-identity elements all other elements of a group G have infinite orders then G is said to be a *mixed group*.

Definition 1.8.4 : An element a of a group (G, *) is said to be *idempotent element* if $a * a = a$.

1.8.1 Illustrative Examples

1. In the group $(G, * = \cdot)$, where $G = \{-1, 1, i, -i\}$ observe that $e = 1$ and

$$(-1)^2 = (-1) \cdot (-1) = 1$$
$$(i)^4 = 1, \ (-i)^4 = 1 \quad \text{and} \quad 1^1 = 1,$$

so that

$$0(-1) = 2, \ 0(i) = 0(-i) = 4 \quad \text{and} \quad 0(1) = 1.$$

Thus in G, each element has a finite order, i.e, G is a torsion group.

2. Consider the group table of the group $(\mathbb{Z}_5, * = +_5)$, where $\mathbb{Z}_5 = \{0, 1, 2, 3, 4\}$:

$* = +_5$	0	1	2	3	4
0	0	1	2	3	4
1	1	2	3	4	0
2	2	3	4	0	1
3	3	4	0	1	2
4	4	0	1	2	3

Observe that here $e = 0$ and

$$0^1 = 0 \Rightarrow 0(0) = 1$$
$$1^5 = \underbrace{1 \cdot 1}_{} * \underbrace{1 \cdot 1}_{} * 1$$
$$= \underbrace{2 \cdot 2}_{} * 1$$
$$= 4 * 1$$
$$= 0 \Rightarrow 0(1) = 5$$

Similarly order of every non-identity element is 5. Notice that order of $\mathbb{Z}_5$ is also 5, which is a prime number. Similarly by making group tables of groups $(\mathbb{Z}_n, +_n)$ when $n = 3, 7$ and 11, we see that order of every element is equal to the order of group. What conclusion will you draw ? Is $(\mathbb{Z}_5, +_5)$ a torsion group ?

3. Each non-identity element of group $(\mathbb{Z}, +)$ is of infinite order because there does not exist any positive integer n such

that $m^n = 0$, $m \in \mathbb{Z}$ where $m^n = m + m + \ldots + m$ (n times). Hence $(\mathbb{Z}, +)$ is a torsion-free group.

4. In the group $(\mathbb{R}^*, \cdot)$, $0(1) = 1$, $0(-1) = 2$ and $0(x) = \infty$, where x is any element of $\mathbb{R}^*$ other than -1 and 1. Hence $(\mathbb{R}^*, \cdot)$ is a mixed group.

5. In the group $(\mathbb{Q}^+, \cdot)$, $e (= 1)$ is the only element of finite order. Hence $(\mathbb{Q}^+, \cdot)$ is a torsion-free group.

6. In the group $(\mathbb{C}^*, \cdot)$ every nth root of unity ($n = 1, 2, \ldots$) is of finite order (see example-1 above for $n = 4$) and the elements $re^{i\theta}$ ($r \neq 1$) are all of infinite orders. Hence $(\mathbb{C}^*, \cdot)$ is a mixed group.

1.8.2 Properties of Order of an Element

Theorem 1.8.1 : The order of every element of a finite group is finite and can not exceed the order of group.

Proof : Let (G, *) be a finite group of order n and $a \in G$. Since * is a b.o., all the following elements

$a^1 = a$, $a^2 = a * a$, $a^3 = a * a * a$, $a^4 = a * a * a * a$, ..., belong to G. But since G is a finite group all of these elements a^1, a^2, a^3, a^4, ... can not be distinct. Thus for $r, s \in \mathbb{Z}^+$ and $r > s$, let $a^r = a^s$ which implies $a^r * a^{-s} = a^s * a^{-s}$ ($a^{-s} = (a^s)^{-1}$ *i.e.*
$a^{r-s} = e$

Since $r - s > 0$ and finite, thus $0(a)$ is finite. Further, let $0(a) = t$ where $t > 0(G) = n$. The elements $a^t = e$, a^1, a^2, a^3, ..., a^{t-1} are all distinct, for, if $a^r = a^s$ $1 \leq s < r \leq t$ then $a^{r-s} = e$ implying that $0(a) = r - s$, which is not possible as $r - s < t$.

Since * is a b.o., all the elements a^1, a^2, ..., a^t (which are distinct and t in number) belong to G which is a contradiction as G has n elements and $t > n$. Thus $t \not> n$ or $0(a) \not> 0(a)$, *i.e.* $0(a) \leq 0(G)$.

Theorem 1.8.2 : Let (G, *) be a group, $a \in G$ and $0(a) = n$. Then :

(i) $0(a) = 0(a^{-1})$

(ii) $a^m = e$ if and only if n divides m.

Binary Operations and Groups 65

(iii) If l is any integer then $0\,(a^l) \leq n$ and $0\,(a^l) = \dfrac{n}{\gcd(n,\,l)}$.

(iv) If p is a positive integer prime to n then $0\,(a^p) = n$.

(v) If $b \in G$ then $0\,(b^{-1}ab) = 0\,(a) = n$ and hence $0\,(ab) = 0\,(ba)$.

(vi) If $l,\,m \in \mathbb{Z}$ then $a^l = a^m$ if and only if $l \equiv m\ (\bmod\ n)$.

(vii) The elements $a^1,\,a^2,\,...,\,a^{n-1},\,a^n = e$ are all distinct.

Proof : Given : $(G,\,*)$ is a group, $a \in G$ and $0\,(a) = n$, i.e. n is least positive integer satisfying $a^n = e$.

(i) $(a^{-1})^n = (a^n)^{-1} = e^{-1} = e \Rightarrow 0\,(a^{-1}) \leq n$.

Let $0 < r < n$ and $0\,(a^{-1}) = r$

then $(a^{-1})^r = e$, or $(a^r)^{-1} = e \Rightarrow a^r = e^{-1} = e$,

which is not possible as $0\,(a) = n$ and $r < n$. Hence such r is not possible, i.e. $0\,(a^{-1}) \not< n$ or $0\,(a^{-1}) = n$. Thus $0\,(a) = 0\,(a^{-1})$.

(ii) Let n divides m. Then there exists $q \in \mathbb{Z}$ such that $m = nq$ and

$$a^m = a^{nq} = (a^n)^q = e^q = e$$

Conversely, let $a^m = e$. Then since $0\,(a) = n$, $m \geq n$ and by division algorithm we can write $m = nq + r$ where $q,\,r \in \mathbb{Z}$ and $0 \leq r < n$. Then $a^m = e \Rightarrow a^{nq+r} = e$

$\Rightarrow a^{nq} * a^r = e \Rightarrow (a^n)^q * a^r = e \Rightarrow e * a^r = e$

$\Rightarrow a^r = e$, which is not possible as $0\,(a) = n$ and $r < n$. Thus $r = 0$ and $m = nq$.

(iii) $(a^l)^n = (a^n)^l = e^l = e$, i.e. $(a^l)^n = e$ which implies that $0\,(a^l) \leq n$.

Let $\gcd(n,\,l) = m$ and $\dfrac{n}{\gcd(n,\,l)} = t$. Then $n = mt$. Let $l = ms$ where $\gcd(t,\,s) = 1$.

Now $(a^l)^q = a^{lq} = e$

$\Leftrightarrow n$ divides lq (by (ii))

$\Leftrightarrow mt$ divides msq

$\Leftrightarrow t$ divides sq

$\Leftrightarrow t$ divides q $(gcd\ (t, s) = 1)$

Thus smallest positive integer q satisfying $(a^l)^q$ is t.

$\Rightarrow 0\ (a^l) = t$, where $t = \dfrac{n}{gcd\ (n, l)}$.

(iv) Let $0\ (a^p) = m$. Since $0\ (a) = n$,
$$a^n = e \Rightarrow (a^n)^p = e^p = e \Rightarrow (a^p)^n = e,$$
which implies that
$$0\ (a^p) \le n,\ or\ m \le n \qquad \ldots(1)$$

Further, since p and n are prime to each other, there exists x, y in $\mathbb{Z}$ such that
$$px + ny = 1$$
$\Rightarrow \qquad a^{px + ny} = a$

$\Rightarrow \qquad a^{px} * a^{ny} = a$

$\Rightarrow \qquad a^{px} = a$

$\Rightarrow \qquad (a^{px})^m = a^m$

$\Rightarrow \qquad (a^p)^{xm} = a^m$

$\Rightarrow \qquad \{(a^p)^m\}^x = a^m$

$\Rightarrow \qquad e^x = a^m \Rightarrow e = a^m$

$\Rightarrow \qquad 0\ (a) \le m \Rightarrow n \le m \qquad \ldots(2)$

By (1) and (2), $\qquad n = m$ or $0\ (a) = 0\ (a^p)$.

(v) Consider $(b^{-1}ab)^2$
$$= (b^{-1}ab) * (b^{-1}ab)$$
$$= b^{-1}a\ (bb^{-1})\ ab \quad \text{(generalised AL)}$$
$$= b^{-1}a^2b$$

Similarly, in general, $(b^{-1}ab)^n = b^{-1}a^nb\ (n \in \mathbb{Z}^+)$

Since $0\ (a) = n \Rightarrow (b^{-1}ab)^n = b^{-1}eb = e$

$\Rightarrow 0\ (b^{-1}ab) \le n$

Take $0 < r < n$, then

$(b^{-1}ab)^r = b^{-1}a^rb \ne e$ as $a^r \ne e$

Binary Operations and Groups 67

$\Rightarrow$ 0 $(b^{-1}ab) \neq r$ or 0 $(b^{-1}ab) \not< n$

Thus $\qquad$ 0 $(b^{-1}ab) = n = 0$ (a).

Further, since
$$a^{-1}(ab)a = (a^{-1}a)ba = ba$$
Thus from above $\quad$ 0 $(ba) = 0$ $\{a^{-1}(ab)a\}$
$$= 0\ (ab).$$
(vi) $\qquad\qquad\qquad a^l = a^m$

$\Leftrightarrow$ $\qquad\qquad\qquad a^{l-m} = e$

$\Leftrightarrow n$ divides $l - m \quad$ (by (ii))

$\Leftrightarrow l \equiv m \pmod{n}$.

SOLVED PROBLEMS SET II

Problem 1.32 : Show that the pair $(\mathbb{R}^*, *)$, where $\mathbb{R}^*$ is set of all real numbers except 0 and the operation $*$ is defined by $a * b = |a| b$, does not form a group.

Solution : (i) Given $*$ is an associative b.o. over $\mathbb{R}^*$, as for any $a, b \in \mathbb{R}^*, a * b = |a|.b \in \mathbb{R}^*$ and for any $a, b, c \in \mathbb{R}^*$.
$$a * (b * c) = a * (|b|c) = |a|(|b|c)$$
$$= (a * b) * c.$$

(ii) In view of Illustrative examples 1.3.1, examples 1 (v) and 3, pair $(\mathbb{R}^*, *)$ does not satisfy either of the two one-sided group axioms, thus $(\mathbb{R}^*, *)$ is not a group.

Problem 1.33 : (a) If a, b, c are the non-identity elements of the Klein's group K_4, find order of the element $(ab^{-1}c)^{-1}$.

(b) In the group $(\mathbb{C}^*, \cdot)$, find order of the element $\dfrac{1+i}{\sqrt{2}}$.

(c) Find order of every element of the group $(\mathbb{Z}_6, +_6)$.

Solution : (a) From the group table of K_4, $b^{-1} = b$ and $a * b = c$, we have by generalised AL
$$ab^{-1}c = abc = cc = e, \text{ so that}$$
$$0\ (ab^{-1}c)^{-1} = 0\ (e^{-1}) = 0\ (e) = 1.$$

(b) For the group $(\mathbb{C}^*, \cdot)$, $e = 1$ and by actual multiplication

$$\left(\frac{1+i}{\sqrt{2}}\right)^8 = 1 \text{ and } \left(\frac{1+i}{\sqrt{2}}\right)^m \neq 1,$$

where $m \in \mathbb{Z}^+$ and $m < 8$.

Thus $\quad 0\left(\dfrac{1+i}{\sqrt{2}}\right) = 8.$

(c) From the table $e = 0$, and

$* = +_6$	0	1	2	3	4	5
0	0	1	2	3	4	5
1	1	2	3	4	5	0
2	2	3	4	5	0	1
3	3	4	5	0	1	2
4	4	5	0	1	2	3
5	5	0	1	2	3	4

Group table for $(\mathbb{Z}_6, +_6)$

$0^1 = 0 \Rightarrow 0\,(0) = 1$

$1^6 = \underbrace{1 * 1} * \underbrace{1 * 1} * \underbrace{1 * 1}$
$ = \underbrace{2 * 2} * 2$
$ = 4 * 2$
$ = 0$

$\Rightarrow \qquad 0\,(1) = 6$

since $\qquad$ Inv $1 = 5$

thus $\qquad 0\,(5) = 6$

as $\qquad 0\,(a) = 0\,(a^{-1})$

$2^3 = \underbrace{2 * 2} * 2$
$ = 4 * 2$
$ = 0$

$\Rightarrow \qquad 0\,(2) = 3 = 0\,(4)$, as Inv $2 = 4$

and $\qquad 3^2 = 3 * 3 = 0 \Rightarrow 0\,(3) = 2$

Binary Operations and Groups 69

Thus in the group $(\mathbb{Z}_6, +_6)$ every element is of finite order and hence $\mathbb{Z}_6$ is a torsion group. In fact every $(\mathbb{Z}_n, +n)$ $(n > 0)$ is a torsion group.

Problem 1.34 : (a) In a group (G, *) if $ba = a^m b^n$, $m, n \in \mathbb{Z}^+$, then show that

$0 (a^m b^{n-2}) = 0 (a^{m-2} b^n) = 0 (ab^{-1})$

(b) If (G, *) is a finite group of even order, show that it has an element $a \neq e$ such that $a * a = e$.

Solution : (a) Since $a^m b^{n-2}$

$$= (a^m b^n) b^{-2}$$
$$= (ba) b^{-2}$$
$$= bab^{-1} b^{-1}$$
$$= b (ab^{-1}) b^{-1}$$
$$= (\bar{b}^1)^{-1} (ab^{-1}) b^{-1},$$

thus $\quad 0 (a^m b^{n-2}) = 0 \{(b^{-1})^{-1} (ab^{-1}) b^{-1}\}$
$$= 0 (ab^{-1}) \quad (\text{as } 0 (b^{-1}ab) = 0 (a)) \quad ...(1)$$

Again $\quad a^{m-2} b^n = a^{-2} (a^m b^n) = a^{-2} (ba) = a^{-2} b (a^{-1} a^2)$
$$= a^{-2} (ba^{-1}) a^2$$

or, $\quad a^{m-2} b^n = (a^2)^{-1} (ba^{-1}) a^2$

so that $\quad 0 (a^{m-2} b^n) = 0 \{(a^2)^{-1} (ba^{-1}) a^2\} = 0 (ba^{-1})$

But $\quad 0 (ba^{-1}) = 0 \{(ba^{-1})^{-1}\} = 0 (ab^{-1})$

thus $\quad 0 (a^{m-2} b^n) = 0 (ab^{-1}) \quad ...(2)$

By (1) and (2)

$0 (a^m b^{n-2}) = 0 (a^{m-2} b^n) = 0 (ab^{-1})$

(b) In a group $e^{-1} = e$ and every non-identity element has its inverse in G. Thus if 0 (G) is even, say 4 and G = {e, a, b, c} then out of 3 non-identity elements atleast one element has to be its own inverse (for example if $a^{-1} = b$ and $b^{-1} = a$ then $c^{-1} = c$).

Thus if G is of even order, atleast one non-identity element, say a, will always be there in G satisfying $a^{-1} = a$, which implies $a * a = e$.

Problem 1.35 : Prove that in a group identity element is the only idempotent element.

Solution : Let $(G, *)$ be a group and $a \in G$. By definition, the element a is idempotent if

$$a * a = a$$
$$\Rightarrow \quad a' * (a * a) = a' * a$$
$$\Rightarrow \quad (a' * a) * a = e$$
$$\Rightarrow \quad e * a = e$$
$$\Rightarrow \quad a = e$$

Thus e is the only idempotent element in G.

Problem 1.36 : (a) In a group $(G, *)$ if $ab = ba$ for $a, b \in G$, $0(a) = n$, $0(b) = m$ such that n and m are relatively prime then prove that $0(ab) = mn$.

(b) Give an example to show that, in general, for any $a, b \in (G, *)$, $0(ab) \neq 0(a) \, 0(b)$.

Solution : (a) Since $0(a) = n$, $0(b) = m$ we have $a^n = e$ and $b^m = e$, $n, m > 0$ and $mn > 0$. Let $0(ab) = k$ so that

$$(ab)^k = e \qquad \ldots(1)$$

Since $ab = ba$, we have (Theorem 1.6.7)

$$(ab)^{mn} = a^{mn} b^{mn}$$
$$= (a^n)^m (b^m)^n$$
$$= e^m e^n$$

or, $(ab)^{mn} = e$ ...(2)

since $0(ab) = k$, by (2)

$$k \leq mn \qquad \ldots(3)$$

By (1) $\qquad a^k = b^{-k}$
$\Rightarrow \qquad a^{mk} = (b^{-k})^m = (b^m)^{-k} = e^{-k} = e$

or, $\qquad a^{mk} = e$

$\Rightarrow n$ divides mk (Theorem 1.8.2. (ii))

$\Rightarrow n$ divides k (as $gcd(n, m) = 1$)

Proceeding same way and considering $b^k = a^{-k}$ by (1), we get

m divides k. Thus mn divides k

$$\Rightarrow \qquad mn \leq k \qquad \qquad \ldots(4)$$

By (3) and (4), $\quad mn = k$.

Thus $\qquad 0\,(ab) = mn$.

(b) Consider the group $G = \{-1, 1, i, -i\}$, $* = \cdot$

Take $\quad a = i$ and $b = -1$ so that

$$0\,(i) = 4 \text{ as } i^4 = 1 \text{ and } 0\,(-1) = 2 \text{ as } (-1)^2 = 1$$

$$0\,(ab) = 0\,(i\,(-1)) = 0\,(-i) = 4, \text{ as } (-i)^4 = 1$$

Thus $\qquad 0\,(ab) \neq 0\,(a)\,0\,(b)$.

1.9 Mappings of Groups

After knowing so much about a single group, one is naturally inclined to 'compare' two or more groups. This comparison is done by studying mappings of groups. Two such important mappings are *homomorphism* and *isomorphism* of groups, which we define now :

Definition 1.9.1 : The map $\sigma : G \to G'$ between any two groups $(G, *)$ and $(G', 0)$ is said to be homomorphism if for any two elements $a, b \in G$.

$$(a * b)\,\sigma = (a\sigma)\,0\,(b\sigma) \qquad \ldots(1)$$

Obviously, since $a, b \in G$ the operation between them will be that of group G, *i.e.* $*$ and since $a\sigma, b\sigma \in G'$ the operation between them will be that of group G', *i.e.* 0. With this understanding

We can rewrite (1) as

$$(ab)\,\sigma = (a\sigma)\,(b\sigma) \qquad \ldots(2)$$

which means that image of 'product' ab is the product of images. Equation (1) (or (2) is called the condition for homomorphism and the map σ appearing there is called the *homomorphic map*.

Note : 1. For brevity we will write *hmp* for homomorphism and *ismp* for isomorphism.

2. Condition (1) also implies that a homomorphic map preserves the group operations.

3. The image set (G) σ is called the *homomorphic image* of G wrt σ. If in addition, σ is onto we say σ an *onto homomorphism* (or *Epimorphism*) and if σ is into, we say σ an *into homomorphism*. Thus for an epimorphism (G) σ = G' and for an into homomorphism (G) σ ⊂ G'.

4. The map $\sigma : G \to G'$ defined by the rule $a\sigma = e'$ for all $a \in G$, where e' is the identity of G', is called the *zero-homomorphism* and the map $I : G \to G$ defined by the rule $aI = a$ for all $a \in G$ is called the *identity-homomorphism*.

5. By repeated application of (2), we notice that if $a_1, a_2, ..., a_n \in G$ then
$$(a_1 a_2 ... a_n)\sigma = (a_1\sigma)(a_2\sigma)...(a_n\sigma),$$
and if $a_1 = a_2 = ... = a_n = a$ (say) we have
$(a^n)\sigma = (a\sigma)^n$ for any positive integer n.

Definition 1.9.2 : A homomorphic map $\sigma : G \to G'$ between any two groups (G, *) and (G', 0) is said to be an *isomorphism* if it is one-one onto as well. Then the group G is said to be *isomorphic* to group G', symbolized as $G \cong G'$.

Thus a mapping $\sigma : G \to G'$ between two groups is an isomorphism if it satisfies following three conditions :

(I) σ is one-to-one mapping, *i.e.*
$x\sigma = y\sigma \Rightarrow x = y$ for any $x, y \in G$.

(S) σ is onto mapping, *i.e.*
(G)σ = G'   (range set = Co-domain) and

(O) σ is homomorphism, *i.e.*
$(xy)\sigma = (x\sigma)(y\sigma), \quad x, y \in G$.

Note on Isomorphism : Isomorphism plays an important role in the study of groups. Groups which are isomorphic can be considered as 'copies' of each other, and isomorphism can be said to be a property of groups which is analogous to that of similarity of sets. For example, consider the group tables of the groups $(\{1, i, -1, -i\}, * = \cdot)$ and $(\{0, 1, 2, 3\}, * = +_4)$ i.e. $\mathbb{Z}_4$:

$* = \cdot$	1	i	-1	$-i$
1	1	i	-1	$-i$
i	i	-1	$-i$	1
-1	-1	$-i$	1	i
$-i$	$-i$	1	i	-1

$* = +_4$	0	1	2	3
0	0	1	2	3
1	1	2	3	0
2	2	3	0	1
3	3	0	1	2

The correspondence between the elements of the two groups is as follows :

$1 \leftrightarrow 0, i \leftrightarrow 1, -1 \leftrightarrow 2$ and $-i \leftrightarrow 3$,

which can be shown to be an isomorphisr.. Thus the group $(\{1, i, -1, -i\}, * = \cdot)$ is isomorphic to the second group $(\{0, 1, 2, 3\}, * = +_4)$. Now, if we re-name the elements 0, 1, 2, 3 respectively as $1, i, -1, -i$, the second group table structurally becomes the same as the first one. For example $3 * 3 = 2$ and when 2 and 3 are re-named as -1 and $-i$ respectively then also, from the first table $(-i) * (-i) = -1$. The 'similarity' of the two groups is due to isomorphism. Similarly we have another isomorphic map between the elements of these two groups, given by :

$1 \leftrightarrow 0, i \leftrightarrow 3, -1 \leftrightarrow 2, -i \leftrightarrow 1.$

In this case the group table of the second group becomes

$* = +_4$	0	3	2	1
0	0	3	2	1
3	3	2	1	0
2	2	1	0	3
1	1	0	3	2

which is also structurally the same as group table one. Can you establish any other isomorphism between these two groups ? (The answer is no).

The advantage of isomorphism is that by making detailed study of a group, informations can be had for all those groups which are isomorphic to it. Homomorphism and isomorphism are Greek words which mean : homo (like), iso (same) and morphism (shape).

Definition 1.9.3 : A property of groups G and G′ which are in isomorphism is said to be *invariant* if G has the property then G′ also has it.

An immediate such property is the order of finite groups. Let G and G′ be finite and G $\cong$ G′ then since there exists a map between G and G′ which is both, one-one and onto thus $0(G) = 0(G')$. Thus isomorphic groups which are finite have the same order. Consequently if G_1 and G_2 are finite groups such that $0(G_1) \neq 0(G_2)$ then G_1 can not be isomorphic to G_2.

We will make a detail study of different mappings of groups in chapter 7. At present let us look at some examples on homomorphism and isomorphism.

1.9.1 Illustrative Examples

1. The map $\sigma : (\mathbb{Z}, +) \to (\mathbb{Z}, +)$ defined by $n\sigma = 4n$, $n \in \mathbb{Z}$, is a hmp as for $n_1, n_2 \in \mathbb{Z}$, "$(n_1 n_2)\sigma$"

$$= (n_1 + n_2)\sigma = 4(n_1 + n_2) = 4n_1 + 4n_2 = n_1\sigma + n_2\sigma$$
$$= ``(n_1\sigma)(n_2\sigma)"$$

In general the map $\sigma : (\mathbb{Z}, +) \to (\mathbb{Z}, +)$ defined by $n\sigma = kn$, where k is some fixed integer, is a homomorphism. Check yourself that this hmp is one-to-one but not onto if $k \neq \pm 1$.

2. Let $\mathbb{C}_x = \{e^{ix} : x \in \mathbb{R}\}$. Then $(\mathbb{C}_{x'}, \cdot)$ is a group of complex numbers of absolute value 1. Define a map $f : (\mathbb{R}, +) \to (\mathbb{C}_{x'}, \cdot)$ by

$$xf = e^{ix} \quad \text{for all } x \in \mathbb{R}$$

Then the map f is an onto hmp (Epimorphism). This map is not one-to-one as

$(x + 2n\pi)f = e^{i(x + 2n\pi)} = e^{ix}$, $n = 0, 1, 2, \ldots$

(The group $(\mathbb{C}_{x'}, \cdot)$ is also denoted by $(0_1, \cdot)$)

3. Let $(G, *)$ be any group. Then the map $g : G \to G$ defined by $ag = a^{-1}$ for all $a \in G$ is one-to-one and onto but it is not hmp as for any $a, b \in G$, "$(ab)\, g$"
$= (a * b)\, g = (a * b)^{-1} = b^{-1} * a^{-1} = bg * ag \neq $"$(ag)(bg)$"

Of course, if the group $(G, *0$ is abelian as well then the map g becomes a hmp.

4. The map $\sigma : (\mathbb{Z}, +) \to (\{-1, 1\}, * = \cdot)$ defined by

$$n\sigma = \begin{cases} 1 \text{ if } n \text{ is zero or even integer} \\ -1 \text{ if } n \text{ is odd integer} \end{cases}$$

is an onto hmp which is not one-to-one.

5. The map $\sigma : (G, *) \to (G', 0)$ defined by $a\sigma = e'$, for all $a \in G$, where e' is identity element of G' is a homomorphism as for any $a, b \in G$, "$(ab)\, \sigma$".
$= (a * b)\, \sigma = e' = e' 0 e' = (a\sigma)\, 0\, (b\sigma) = $"$(a\sigma)(b\sigma)$"

This hmp (called zero-hmp) is neither one-to-one nor onto.

6. The identity hmp $I : (G, *) \to (G, *)$ defined by $aI = a$ for all $a \in G$ is an isomorphism as for any $a, b \in G$, "$(ab)\, I$"
$= (a * b)\, I = a * b = (a)\, I * (b)\, I = (aI)(bI)$.

The map I is one-to-one as $aI = bI \Rightarrow a = b$ and the map I is onto as well as $(G)\, I = G$. thus all the three conditions I, S and O for an isomorphism is satisfied by the map I, hence $G \cong G$ for every group G, i.e. any group G is isomorphic to itself.

7. Here is an example of a set being isomorphic to its subset. Consider the map $\sigma : (\mathbb{Z}, +) \to (2\mathbb{Z}, +)$ defined by $n\sigma = \pm 2n$, for all $n \in \mathbb{Z}$. Then σ is an isomorphism (check the conditions I, S and O for this σ) hence $\mathbb{Z} \cong 2\mathbb{Z}$, i.e. the set of integers $\mathbb{Z}$ (forming a group under $* = +$) is isomorphic to its subset $2\mathbb{Z}$ of even integers (forming a group under $* = +$).

8. The map $\sigma : (\mathbb{Z}, +) \to (\{1, 3^{\pm 1}, 3^{\pm 2}, \ldots\}, * = \cdot)$ defined by $n\sigma = 3^n$ for all $n \in \mathbb{Z}$ is an isomorphism. Thus the infinite group $(\mathbb{Z}, +)$ is isomorphic to the infinite group $(\{3^n : n \in \mathbb{Z}\}, * = \cdot)$.

9. The map $\sigma : (\mathbb{C}, +) \to (\mathbb{C}, +)$ defined by $(a + ib)\sigma = a - ib$ is an isomorphism and hence $\mathbb{C} \cong \mathbb{C}$.

10. The map $i_a : (G, *) \to (G, *)$ defined by
$$xi_a = a^{-1}xa, \text{ for all } x \in G,$$
where a is a fixed element of G and $a^{-1}xa$ stands for $a^{-1} * x * a$, is an isomorphism. (For details, see Theorem 7.5.2 in chapter 7).

Similarly the map $\bar{i}_a : (G, *) \to (G, *)$ defined by
$$x\bar{i}_a = ax\bar{a}^1 \quad \text{for all } x \in G,$$
where a is a fixed element of G, is an isomorphism.

Solved Problems Set–III

Problem 1.37 : (a) Prove that the map $\sigma : (\mathbb{R}, +) \to (\mathbb{R}^+, \cdot)$ defined by $x\sigma = e^x$ for all $x \in \mathbb{R}$, is an isomorphism.

(b) Prove that the map $\tau : (\mathbb{R}^+, \cdot) \to (\mathbb{R}, +)$ defined by $x\tau = \log x$ for all $x \in \mathbb{R}^+$ is an isomorphism.

Solution : (a) Consider the map $\sigma : (\mathbb{R}, +) \to (\mathbb{R}^+, \cdot)$ defined as $x\sigma = e^x \; \forall \; x \in \mathbb{R}$. We will show that the map σ satisfies all the 3 conditions for an isomorphism :

(I) Map σ is one-to one as for any $x, y \in \mathbb{R}$,
$$x\sigma = y\sigma$$
$$\Rightarrow \quad e^x = e^y$$
$$\Rightarrow \quad x = y$$

(S) Map σ is onto as for any $r \in \mathbb{R}^+$ we have $\log r \in \mathbb{R}$ such that
$$(\log r)\sigma = e^{\log r} = r, \text{ and finally}$$

(O) for $\; x, y \in \mathbb{R}$,
$$\text{``}(xy)\sigma\text{''}$$
$$= (x + y)\sigma$$
$$= e^{x+y} = e^x \cdot e^y = (x\sigma) \cdot (y\sigma) = \text{``}(x\sigma)(y\sigma)\text{''}$$

Thus $\quad \mathbb{R} \cong \mathbb{R}^+$

(b) Proceeding exactly the same way, we can show that the map τ is an isomorphism, so that $\mathbb{R}^+ \cong \mathbb{R}$.

Problem 1.38: If $\sigma : G \to G'$ is an isomorphism, prove that $\sigma^{-1} : G' \to G$ is also an isomorphism.

Solution: Since $\sigma : G \to G'$ is one-to-one onto hence σ^{-1} is defined, and $\sigma^{-1} : G' \to G$ is itself one-to-one and onto. (Prove it). Further, let $a, b \in G'$. and $x = a\sigma^{-1}$ and $y = b\sigma^{-1}$, then $a = x\sigma$ and $b = y\sigma$; $x, y \in G$.

Now $\qquad ab = (x\sigma)(y\sigma)$
$\qquad\qquad\quad = (xy)\sigma \quad$ (σ is hmp),

thus $\qquad (ab)\sigma^{-1} = xy = (a\sigma^{-1})(b\sigma^{-1})$, which is the condition for homomorphism for the inverse map σ^{-1}. Thus $G \cong G' \Rightarrow G' \cong G$.

Problem 1.39: If $\sigma : G \to G'$ and $\tau : G' \to G''$ are isomorphisms, where G, G', G'' are groups under their respective binary operations, then prove that the composite mapping $\sigma\tau : G \to G''$ is an isomorphism.

Solution: $\sigma\tau$ is one-to-one : If x, y are any two distinct elements of G then, since σ is one-to-one, $x\sigma$ and $y\sigma$ are distinct elements of G' and hence $(x\sigma)\tau$ and $(y\sigma)\tau$ are distinct elements of G'' as τ is one-to-one. Thus $x(\sigma\tau)$ and $y(\sigma\tau)$ are distinct whenever x and y are distinct implying that $\sigma\tau$ is one-to-one, which is condition (I) for an isomorphism.

$\sigma\tau$ is onto : Take an element x'' of G''.

Since τ is onto, there exists some x' in G' such that $x'\tau = x''$. Also σ is onto, there exists some $x \in G$ such that $x\sigma = x'$. Thus

$$x(\sigma\tau) = (x\sigma)\tau = x'\tau = x''$$

showing that for each x'' in G'', there exists some x in G such that $x(\sigma\tau) = x''$, i.e. the map $\sigma\tau$ is onto, which is condition (S) for an isomorphism.

$\sigma\tau$ is a homomorphism : For any $x, y \in G$,

$\qquad (xy)(\sigma\tau)$

$\qquad\qquad = \{(xy)\sigma\}\tau$

$\qquad\qquad = \{(x\sigma)(y\sigma)\}\tau \quad$ (as σ is hmp)

$\qquad\qquad = \{(x\sigma)\tau\}\{(y\sigma)\tau\} \quad$ (as τ is hmp)

$\qquad\qquad = \{x(\sigma\tau)\}\{y(\sigma\tau)\},$

which is condition (O) for an isomorphism. Thus the map $\sigma\tau$: $G \to G''$ is an isomorphism.

Problem 1.40 : Prove that the relation $\cong$ of being isomorphic in the set $\mathcal{G}$ of all groups is an equivalence relation.

Solution : Let $\mathcal{G} = \{G, G', G'', \ldots \}$ be the set of all groups.

(i) Relation $\cong$ is reflexive : In Illustrative examples 1.9.1, we shown that the identity map $I : G \to G$, for any group G, is an isomorphism (see Ex. 6) so that $G \cong G$, *i.e.* $\cong$ is reflexive.

(ii) Relation $\cong$ is symmetric : By problem 1.38, if $\sigma : G \to G'$ is an isomorphism then $\sigma^{-1} : G' \to G$ is also an isomorphism so that if $G \cong G'$ then $G' \cong G$, *i.e.* $\cong$ is symmetric.

(iii) Relation $\cong$ is transitive : Let $\sigma : G \to G'$ and $\tau : G' \to G''$ be isomorphisms then, as shown in problem 1.39, the composite map $\sigma\tau : G \to G''$ is an isomorphism. Thus if $G \cong G'$ and $G' \cong G''$ then $G \cong G''$, *i.e.* $\cong$ is transitive.

By (i), (ii) and (iii) the relation $\cong$ of being isomorphic is an equivalence relation in $\mathcal{G}$.

Note : According to general theory of equivalence relation (See Prelim. 2 (iv)), the equivalence relation $\cong$ will partition the set $\mathcal{G}$ of all groups into mutually disjoint equivalence classes, each class containing groups isomorphic to each other. Thus, structurally, all the groups belonging to same equivalence class will be identical.

Problem 1.41 : Show that the map $\sigma : (\mathbb{Z}, *) \to (\mathbb{Z}, 0)$ defined by $n\sigma = -n$, for all $n \in \mathbb{Z}$, is an isomorphism where the binary operations $*$ and 0 are defined respectively as

$m * n = m + n + mn$ and $m o n = m + n - mn$, for all $m, n \in \mathbb{Z}$.

Solution : For any $x, y \in \mathbb{Z}$,

"$(xy)\sigma$"

$= (x * y)\sigma$

$= -(x * y) = -(x + y + xy) = -x - y - xy$

$= (x\sigma) + (y\sigma) - (x\sigma)(y\sigma)$

$= (x\sigma) \, 0 \, (y\sigma)$

$= "(x\sigma)(y\sigma)"$

Showing that σ is a hmp. Check that σ is one-to-one and onto.

Problem 1.42 : Show that the group of nth roots of unity is isomorphic to the group $(\mathbb{Z}_n, +n)$ where $\mathbb{Z}_n = \{0, 1, 2, ..., n-1\}$.

Solution : Let G be the group of nth roots of unity under $* = \cdot$. Then set $G = \{e^{i2r\pi/n} : r = 0, 1, 2, ..., (n-1)\}$ (See Problem 1.12). Define a map $\sigma : G \to \mathbb{Z}_n$ by

$$(e^{i2r\pi/n})\sigma = r, \quad r = 0, 1, 2, ..., n-1$$

and show that σ satisfies all three conditions of isomorphism. Thus $G \cong \mathbb{Z}_n$.

1.10 Direct Product of Groups

Another property which is associated with more than one group is the direct products of two groups. Unlike the case of group of n tuples of real numbers (see Illustrative examples 1.4.1, example A (6)), in a direct product we build a new group from two given groups $(G_1, *)$ and $(G_2, 0)$ by considering their Cartesian product $G_1 \times G_2$ of all ordered pairs of elements of G_1 and G_2. (In particular case the groups G_1 and G_2 may be the same). We construct this group in the form of following theorem.

Theorem 1.10.1 : Let $(G_1, *)$ and $(G_2, 0)$ be two groups. Then their Cartesian product $G_1 \times G_2 = \{(g_1, g_2) : g_1 \in G_1$ and $g_2 \in G_2\}$ form a group under the operation × defined by

$$(g_1, g_2) \times (h_1, h_2) = (g_1 * h_1, g_2 \, 0 \, h_2),$$

for all $g_1, h_1 \in G_1$ and $g_2, h_2 \in G_2$.

Proof : We will verify all the group axioms for the pair $G_1 \times G_2$ and × :

(i) Since $g_1 * h_1 \in G_1$ and $g_2 \, 0 \, h_2 \in G_2$,

thus $(g_1 * h_1, g_2 0 h_2) \in G_1 \times G_2$, i.e. the operation × is a b.o. over $G_1 \times G_2$.

(ii) Using juxtaposition, we have

$$(g_1, g_2) [(h_1, h_2) (k_1, k_2)]$$
$$= (g_1, g_2) [(h_1 k_1, h_2 k_2)]$$

$$= (g_1(h_1k_1), g_2(h_2k_2))$$
$$= ((g_1h_1)k_1, (g_2h_2)k_2)$$
$$= [(g_1, g_2)(h_1, h_2)](k_1, k_2)$$

showing that × satisfies AL.

(iii) Let e_1 and e_2 denote identity elements of groups G_1 and G_2 respectively, then
$$(g_1, g_2)(e_1, e_2) = (g_1e_1, g_2e_2) = (g_1, g_2)$$
$$= (e_1g_1, e_2g_2)$$
$$= (e_1, e_2)(g_1, g_2)$$

i.e. (e_1, e_2) is identity element in $G_1 \times G_2$.

(iv) Let g_1^{-1} = Inv of g_1 and g_2^{-1} = Inv of g_2 then
$$(g_1, g_2)(g_1^{-1}, g_2^{-1}) = (g_1g_1^{-1}, g_2g_2^{-1}) = (e_1, e_2)$$
$$= (g_1^{-1}g_1, g_2^{-1}g_2)$$
$$= (g_1^{-1}, g_2^{-1})(g_1, g_2)$$

i.e. (g_1^{-1}, g_2^{-1}) is the inverse element of (g_1, g_2) in $G_1 \times G_2$.

Thus $G_1 \times G_2$ is a group under the operation × defined above. This group is called the *direct product* of the groups G_1 and G_2.

Note : 1. If, for example * = + and 0 = · then the b.o. × over $G_1 \times G_2$ will be defined by
$$(g_1, g_2) \times (h_1, h_2) = (g_1 + h_1, g_2 \cdot h_2).$$

2. If, in addition, both the groups $(G_1, *)$ and $(G_2, 0)$ are abelian then their direct product $(G_1 \times G_2, \times)$ is also abelian.

3. If one of the groups G_1 or G_2 is of infinite order then $0(G_1 \times G_2) = \infty$, and if both the groups are of finite orders, say $0(G_1) = m_1$ and $0(G_2) = m_2$, then $0(G_1 \times G_2) = m_1m_2$.

4. In the direct product $G_1 \times G_2$, the two groups may be same. For example, if $G_1 = G_2 = (\mathbb{R}, +)$ then their direct product $G_1 \times G_2$ is the group $(\mathbb{R}^2, +)$, where $\mathbb{R}^2 = \mathbb{R} \times \mathbb{R}$ is the Euclidean plane.

Review Exercise on Chapter 1

1. Test if the following binary operations defined over $\mathbb{Z}$ satisfy commutative and associative laws :

(i) $a * b = a$

(ii) $a * b$ = smaller of a or b (or the common value if $a = b$)

(iii) $a *' b = (a * b) + 2$, where $*$ is defined as in (ii), and

(iv) $a * b = 2^{ab}$

where $a, b \in \mathbb{Z}$.

2. (i) On $\mathbb{Z}^+$ define $*$ as $a * b = c$, where c is atleast 5 less than ab. Is $*$ a b.o. over $\mathbb{Z}^+$?

(ii) On $\mathbb{Z}^+$ define $*$ as $a * b = c$, where c is smallest integer greater than both, a and b. Is $*$ an associative b.o. ?

3. (a) Over the set $\mathbb{R}^*$, show the following :

(i) The b.o. $*$ defined by $a * b = \min \{a, b\}$, $a, b \in \mathbb{R}^*$, satisfies both commutative and associative laws.

(ii) The b.o. $*'$ defined by $a\,'\,b = |\,a\,|\,b$, $a, b \in \mathbb{R}^*$, satisfies associative law but not the commutative law.

(iii) The b.o. $*''$ defined by $a *'' b = |\,a - b\,|$ satisfies commutative law but not the associative law.

(b) Show that the b.o. o defined over $\mathbb{Q}^*$ by the rule $poq = \dfrac{p}{q}$, $p, q \in \mathbb{Q}^*$, is neither associative nor commutative.

4. Explain why the following pairs of non-empty set and b.o. do not form a group.

(i) Set $\mathbb{R}$ and $* = \cdot$ (multiplication),

(ii) Set $\mathbb{R}$ and $* = -$ (subtraction),

(iii) The power set $p\,(X)$ of all subsets of a non-empty set X and $* =$ union of A, B where A, B $\in p\,(X)$.

5. Find the identity element e and the inverse of an element a in the case of the following groups :

(i) $(\mathbb{R} - \{-1\}, *)$ where $a * b = a + b + ab$, for all $a, b \in \mathbb{R} - \{-1\}$.

(ii) $\left(\mathbb{R} - \left\{\dfrac{-1}{2}\right\}, *\right)$ where $a * b = a + b + 2ab$, $a, b \in \mathbb{R} - \{-1/2\}$

(iii) ($\mathbb{R}^*$, 0) where $a 0 b = 10ab$, $a, b \in \mathbb{R}^*$.

(iv) ($\mathbb{Z}$, *) where $m * n = m + n - 1$, $m, n \in \mathbb{Z}$.

6. In a semi group G, prove the following

(i) $\quad\quad\quad\quad\quad e_L = e_R$, and

(ii) $\quad\quad\quad\quad\quad a'_L = a'_R$

where e_L, e_R are left and right identity elements and a'_L, a'_R are left and right inverse elements of a.

(iii) Give an example to show that in a groupoid $a'_L \neq a'_R$.

7. Show that the following pairs form a group :

(a) $A = \{z \in \mathbb{C} : z^n = 1$ where n is fixed arbitrary positive integer$\}$, $* = \cdot$

(b) $\mathbb{C}_x = \{e^{ix} : x \in \mathbb{R}\}$, $* = \cdot$

8. (i) If $a, b \in G$, where (G, *) is a group such that $a^{-1} * b^2 * a = b * a$, prove that $b = a$.

(ii) In a group G, prove that $0\,(ab^{-1}) = 0\,(a^{-1}b) = 0\,(ba^{-1})$

9. Let $0\,(G) = n$ and p be a prime number which divides n then prove that G has an element of order p.

10. Let G be a finite abelian group with $0\,(G) = n$ and p and n be relatively prime to each other then prove that there exists some $x \in G$ such that $a = x^p\ \forall\ a \in G$.

11. Let M be the general linear group $GL_2\,(\mathbb{R})$ of all 2×2 matrices $\begin{bmatrix} a & b \\ c & d \end{bmatrix}$ such that $ad - bc \neq 0$ under the b.o. of matrix multiplication, then prove that the map $\sigma : M \to \mathbb{R}^*$, where ($\mathbb{R}^*, \cdot$) is group of non-zero real numbers, defined by

$$\begin{bmatrix} a & b \\ c & d \end{bmatrix} \sigma = ad - bc$$

is an onto hmp.

12. Show the following :

(i) ($\mathbb{Q}$, +) is not isomorphic to ($\mathbb{R}$, +).

(ii) ($\mathbb{Z}$, +) is not isomorphic to ($\mathbb{R}$, +).

2 PERMUTATION AND GROUPS OF PERMUTATION

2.1 Permutation over a finite set

Consider a finite set A = {x, y, z} having any three elements. A permutation over A is a rearrangement of positions of its elements. For example, one way to rearrange the elements of A is to interchange the positions of first two elements x and y and to keep the third element at its own position so that we can write the set A = {y, x, z}. We then say that x is carried over to y, y is carried over to x and is to z itself, i.e., $x \to y$, $y \to x$ and $z \to z$. Before and after this rearrangement there is no change in the set A except for the positions of its elements. Likewise we can rearrange the positions of the three elements of A and obviously there will be total $\lfloor 3$ such rearrangements. If we look at this situation in terms of a function, we can say that A is the domain and A itself is the co-domain of the rule (function) of rearrangement of elements of A and thus we define a permutation over a finite set as follows :

Definition 2.1.1 : A permutation over a finite set A is a function $f : A \to A$ which is both, one to one and onto.

2.1.1 Notations

(1) A permutation over a finite set is denoted with help of two rows put in a small bracket. The elements of the set are written along the first row and the rearranged elements are written along the second row and the notation is called a *two-row notation*. Thus we write a permutation as

$$\begin{pmatrix} \to \text{elements of the set} \leftarrow \\ \to \text{elements in rearranged form} \leftarrow \end{pmatrix}$$

For example, all the $\lfloor 3$ permutations over above set $A = \{x, y, z\}$ will be written as :

$$\begin{pmatrix} x & y & z \\ x & y & z \end{pmatrix}, \begin{pmatrix} x & y & z \\ x & z & y \end{pmatrix}, \begin{pmatrix} x & y & z \\ y & x & z \end{pmatrix}$$

$$\begin{pmatrix} x & y & z \\ y & z & x \end{pmatrix}, \begin{pmatrix} x & y & z \\ z & x & y \end{pmatrix}, \begin{pmatrix} x & y & z \\ z & y & x \end{pmatrix}$$

Notice that the first permutation in which the two rows are identical is one of the $\lfloor 3$ permutations over set A.

(2) *The set of all permutations* over a finite set A will be denoted by S_A. Thus if we denote above six permutations over set $A = \{x, y, z\}$ respectively by $\alpha_1, \alpha_2, \alpha_3, \alpha_4, \alpha_5$ and α_6, i.e., $\alpha_1 = \begin{pmatrix} x & y & z \\ x & y & z \end{pmatrix}, \alpha_2 = \begin{pmatrix} x & y & z \\ x & z & y \end{pmatrix}$, etc then the set of all permutations over A will be $S_A = \{\alpha_1, \alpha_2, \alpha_3, \alpha_4, \alpha_5, \alpha_6\}$. Notice that if $0(A) = 3$ then $0(S_A) = \lfloor 3$. Similarly a set B of any four elements can be rearranged in $\lfloor 4$ ways so that when $0(B) = 4$, $O(S_B) = \lfloor 3$. In general, S_X, the set of all permutations over a finite set $X = \{a_1, a_2, ... a_n\}$ having n elements will contain $\lfloor n$ elements. Thus if $0(X) = n$ then $0(S_X) = \lfloor n$.

(3) *The set S_n* : In particular if elements of above set X are the first n-natural numbers 1, 2, 3, ... n, (in place of any n elements) then the set S_X of all permutations over X will be denoted by S_n, where the suffix n in S_n denotes the total number of natural numbers starting from 1, being rearranged (permutated).

Thus, if $X = \{1, 2, 3\}$ the set S_x will be denoted by S_3; if $X = \{1, 2, ,3 \ 4\}$ the set S_x will be denoted by S_4 and, in general, if $X = \{1, 2, 3, ...n\}$, the set S_x will be denoted by S_n.

For example when $X = \{1, 2, 3\}$, the set of all permutations over X will be

S_3 = set of all permutations over first 3 natural numbers

$$= \left\{ \begin{pmatrix} 1 & 2 & 3 \\ 1 & 2 & 3 \end{pmatrix}, \begin{pmatrix} 1 & 2 & 3 \\ 1 & 3 & 2 \end{pmatrix}, \begin{pmatrix} 1 & 2 & 3 \\ 2 & 1 & 3 \end{pmatrix}, \begin{pmatrix} 1 & 2 & 3 \\ 2 & 3 & 1 \end{pmatrix}, \right.$$

$$\begin{pmatrix} 1 & 2 & 3 \\ 3 & 1 & 2 \end{pmatrix}, \begin{pmatrix} 1 & 2 & 3 \\ 3 & 2 & 1 \end{pmatrix} \Big\},$$

so that $0(S_3) = \lfloor 3$. Similarly $0(S_4) = \lfloor 4$, ..., in general $0(S_n) = \lfloor n$.

We will see in the subsequent sections that the set S_n plays a significant role in studying the symmentaries of a number of geometrical figures.

2.1.2 A b.o. over the set S_x

We have seen in chapter 1 that the set B_A of all one-to-one mappings from a set A onto itself from a group under the b.o. as composition of maps. In view of definition 2.1.1, elements of S_x, where X is set of any n elements are one to one maps of X onto X itself. This motivates us to look for a b.o. over the set S_x and to construct a possible group out of this pair.

Definition 2.1.2 : (Product of two permutations) : The product of any two permutations $\alpha_1, \alpha_2 \in S_x$ is the composite of the two maps $\alpha_2 \, 0 \, \alpha_1$ which is also a one-one onto map from, X to X.

The composite map $\alpha_2 \, 0 \, \alpha_1$ is denoted by $\alpha_1 \alpha_2$ and hence the product of $\alpha_1, \alpha_2 \in S_x$ will be $\alpha_1 \alpha_2$, where $\alpha_1 \alpha_2 \in S_x$ so the *product of two permutations is a b.o. over set* S_x.

Method to find the product of permutations : Let $0(X) = 4$, and $X = \{x, y, z, t\}$ so that $0(S_x) = \lfloor 4$. Take any three elements α, β, γ from the set S_x and let

$$\alpha = \begin{pmatrix} x & y & z & t \\ y & x & z & t \end{pmatrix}, \beta = \begin{pmatrix} x & y & z & t \\ x & z & y & t \end{pmatrix}$$

and $\gamma = \begin{pmatrix} x & y & z & t \\ x & y & t & z \end{pmatrix}$.

In α, $x \to y$ and in β, $y \to z$ so in the product (composite map) $\alpha\beta$, x is carried over to z. Thus we have the following diagram

$$x \xrightarrow{\alpha} y \xrightarrow{\beta} z$$
$$\underbrace{}_{\alpha\beta}$$

Similarly in α, $y \to x$, and in β, $x \to x$ so that in $\alpha\beta$, $y \to x$ and so on and the product permutation $\alpha\beta$ is given by

$$\alpha\beta = \begin{pmatrix} x & y & z & t \\ z & x & y & t \end{pmatrix}$$

Likewise,
$$\beta\alpha = \begin{pmatrix} x & y & z & t \\ y & z & x & t \end{pmatrix}$$

Both, $\alpha\beta$ and $\beta\alpha$ are one of the total $\underline{|4}$ elements of the set S_x so that $\alpha\beta$, $\beta\alpha \in S_x$ where $\alpha, \beta \in S_x$.

2.1.3 Properties of elements of S_x

(1) *Equality of two permutations* : In a permutation we can interchange any two columns without affecting the permutation. The permutation α given above is equal to

$$\begin{pmatrix} y & x & t & z \\ x & y & t & z \end{pmatrix}, \text{ i.e.,}$$

$$\begin{pmatrix} x & y & z & t \\ y & x & z & t \end{pmatrix} = \begin{pmatrix} y & x & z & t \\ x & y & z & t \end{pmatrix}$$

Notice that in permutation given in left side, $x \to y$, $y \to x$, $z \to z$ and $t \to t$ and in permutation given in right side also $x \to y$, $y \to x$, $z \to z$, $t \to t$ and hence they are the same (maps). Similarly

$$\begin{pmatrix} x & y & z & t \\ t & z & y & x \end{pmatrix} = \begin{pmatrix} y & t & x & z \\ z & x & t & y \end{pmatrix},$$

$$\begin{pmatrix} 1 & 2 & 3 & 4 \\ 3 & 1 & 4 & 2 \end{pmatrix} = \begin{pmatrix} 3 & 2 & 4 & 1 \\ 4 & 1 & 2 & 3 \end{pmatrix}, \text{ etc.}$$

Note that in 2.1.2,

$$\alpha\beta = \begin{pmatrix} x & y & z & t \\ z & x & y & t \end{pmatrix} \ne \begin{pmatrix} x & y & z & t \\ y & z & x & t \end{pmatrix} = \beta\alpha$$

(2) A permutation can be multiplied any number of times with itself to give another permutation of the same set. For example if $\sigma = \begin{pmatrix} x & y & z & t \\ z & t & y & x \end{pmatrix} \in S_x$ where $X = \{x, y, z, t\}$ then

$$\sigma^2 = \sigma \cdot \sigma = \begin{pmatrix} x & y & z & t \\ y & x & t & z \end{pmatrix},$$

$$\sigma^3 = \sigma^2 \cdot \sigma = \begin{pmatrix} x & y & z & t \\ y & x & t & z \end{pmatrix} \begin{pmatrix} x & y & z & t \\ z & t & y & x \end{pmatrix} = \begin{pmatrix} x & y & z & t \\ t & z & x & y \end{pmatrix} = \sigma \cdot \sigma^2$$

$$\sigma^4 = (\sigma^2)^2 = \sigma^2 \cdot \sigma^2 = \begin{pmatrix} x & y & z & t \\ x & y & z & t \end{pmatrix}, \text{etc., so that if } \sigma \in S_x \text{ then}$$
$\sigma^2, \alpha^3, \sigma^4, \ldots \in S_x$.

Based on this we define *order* of a permutation as follows :

Definition 2.1.3: If $\alpha \in S_x$ such that both the rows are identical in α then α is denoted by I.

Thus if $X = \{a_1, a_2, \ldots a_n\}$
then I in S_x is given by

$$I = \begin{pmatrix} a_1 & a_2 & \ldots & a_n \\ a_1 & a_2 & \ldots & a_n \end{pmatrix}$$

Similarly if $X = \{1, 2, \ldots n\}$ then I in S_n is given by

$$I = \begin{pmatrix} 1 & 2 & \ldots & n \\ 1 & 2 & \ldots & n \end{pmatrix}.$$

Definition 2.1.4 : If $\alpha \in S_x$, where X is any finite set, such that $\alpha^n = I$, where n is least such positive integer then n is called the *order of permutation* α, denoted by $0(\alpha)$.

Thus if $0(\alpha) = n$ and m is any positive integer such that $\alpha^m = I$ then $m \geq n$.

(3) *Commutativity and associativity of permutation product* : We have seen in 2.1.2 that $\alpha\beta \neq \beta\alpha$ where $\alpha, \beta \in S_x$ where $X = \{x, y, z, t\}$. In general, if X has n elements and $\alpha, \beta \in S_x$ then $\alpha\beta \neq \beta\alpha$. so that permutation product does not obey commutative law.

For α, β and γ as given in 2.1.2 let us calculate the products $\alpha(\beta\gamma)$ and $(\alpha\beta)\gamma$.

$$\alpha\,(\beta\gamma) = \begin{pmatrix} x & y & z & t \\ y & x & z & t \end{pmatrix} \left\{ \begin{pmatrix} x & y & z & t \\ x & z & y & t \end{pmatrix} \begin{pmatrix} x & y & z & t \\ x & y & t & z \end{pmatrix} \right\}$$

$$= \begin{pmatrix} x & y & z & t \\ y & x & z & t \end{pmatrix} \left\{ \begin{pmatrix} x & y & z & t \\ x & t & y & z \end{pmatrix} \right\} = \begin{pmatrix} x & y & z & t \\ t & x & y & z \end{pmatrix}$$

and

$$(\alpha\beta)\,\gamma = \left\{ \begin{pmatrix} x & y & z & t \\ y & x & z & t \end{pmatrix} \begin{pmatrix} x & y & z & t \\ x & z & y & t \end{pmatrix} \right\} \begin{pmatrix} x & y & z & t \\ x & y & t & z \end{pmatrix}$$

$$= \left\{ \begin{pmatrix} x & y & z & t \\ z & x & y & t \end{pmatrix} \right\} \begin{pmatrix} x & y & z & t \\ x & y & t & z \end{pmatrix} = \begin{pmatrix} x & y & z & t \\ t & x & y & z \end{pmatrix} = \alpha\,(\beta\gamma)$$

Thus for any $\alpha, \beta, \gamma \in S_x$ when $0(X) = 4$, $(\alpha\beta)\,\gamma = \alpha\,(\beta\gamma)$ showing that α, β, γ satisfy associative law. In general, if X has n elements and $\alpha, \beta, \gamma \in S_x$ then also permutation product satisfy associative law.

Thus permutation product obeys AL but not the CL.

(4) *Identity and inverse permutation* : The permutation I, as described in the definition 2.1.3, in which both the rows are identical serves as an identify element under permutation product. Thus when $S = \{x\ y\ z\ t\}$, by definition 2.1.3

$$I = \begin{pmatrix} x & y & z & t \\ x & y & z & t \end{pmatrix} \text{ and if } \alpha \in S_x \text{ where}$$

$$\alpha = \begin{pmatrix} x & y & z & t \\ y & x & z & t \end{pmatrix},$$

$$I\alpha = \begin{pmatrix} x & y & z & t \\ y & x & z & t \end{pmatrix} \text{ and } \alpha I = \begin{pmatrix} x & y & z & t \\ y & x & z & t \end{pmatrix}$$

so that $I\alpha = \alpha I = \alpha$

which shows that the permutation I is the identity element of the set S_x when $X = \{x\ y\ z\ t\}$ under permutation product. In general when X has n elements the permutation I having both the rows identical in it serves as identity element in S_x. The permutation I is called the *Identity permutation*.

Permutation and Groups of Permutation

The inverse of a permutation is obtained by interchanging the two rows of the permutation because the newly obtained permutation satisfies the definition of an inverse element. Thus if $v \in S_x$ where $X = \{x\ y\ z\ t\}$ and $v = \begin{pmatrix} x & y & z & t \\ z & t & y & x \end{pmatrix}$

We define *inverse of v*,

$$v^{-1} = \begin{pmatrix} z & t & y & x \\ x & y & z & t \end{pmatrix} \text{ which by}$$

equality of two permutations is

$$v^{-1} = \begin{pmatrix} x & y & z & t \\ t & z & x & y \end{pmatrix}.$$

Now, $vv^{-1} = \begin{pmatrix} x & y & z & t \\ x & y & z & t \end{pmatrix} = v^{-1}v$ so that $vv^{-1} = v^{-1}v = I$, showing that the permutation v^{-1} as obtained above is the inverse of the permutation v. Notice that when $v \in S_x$, v^{-1} also $\in S_x$ where $X = \{x\ y\ z\ t\}$.

In general we define identity permutation and inverse of a permutation as follows :

Definition 2.1.5 : A permutation over a finite set having n elements in which both the rows are identical is the *identity permutation*, denoted by I.

Definition 2.1.6 : *The inverse of a permutation α over a finite set having n elements is the one which is obtained by interchanging the two rows of the permutation and is denoted by α^{-1}.*

Notice that in both the definitions, operation is the permutation product and both, I and $\alpha^{-1} \in S_x$ for any finite set X.

2.1.4 Illustrative Examples

(1) Let $\alpha, \beta \in S_7$ and

$$\alpha = \begin{pmatrix} 1 & 2 & 3 & 4 & 5 & 6 & 7 \\ 7 & 6 & 5 & 4 & 3 & 2 & 1 \end{pmatrix} \text{ and } \beta = \begin{pmatrix} 1 & 2 & 3 & 4 & 5 & 6 & 7 \\ 7 & 6 & 5 & 4 & 2 & 1 & 3 \end{pmatrix}$$

then
$$\alpha\beta = \begin{pmatrix} 1 & 2 & 3 & 4 & 5 & 6 & 7 \\ 3 & 1 & 2 & 4 & 5 & 6 & 7 \end{pmatrix},$$

$$\beta\alpha = \begin{pmatrix} 1 & 2 & 3 & 4 & 5 & 6 & 7 \\ 1 & 2 & 3 & 4 & 6 & 7 & 5 \end{pmatrix},$$

$$(\alpha\beta)^{-1} = \begin{pmatrix} 3 & 1 & 2 & 4 & 5 & 6 & 7 \\ 1 & 2 & 3 & 4 & 5 & 6 & 7 \end{pmatrix} = \begin{pmatrix} 1 & 2 & 3 & 4 & 5 & 6 & 7 \\ 2 & 3 & 1 & 4 & 5 & 6 & 7 \end{pmatrix}$$

$$(\beta\alpha)^{-1} = \begin{pmatrix} 1 & 2 & 3 & 4 & 6 & 7 & 5 \\ 1 & 2 & 3 & 4 & 5 & 6 & 7 \end{pmatrix} = \begin{pmatrix} 1 & 2 & 3 & 4 & 5 & 6 & 7 \\ 1 & 2 & 3 & 4 & 7 & 5 & 6 \end{pmatrix}$$

and

$$\beta^{-1}\alpha^{-1} = \begin{pmatrix} 1 & 2 & 3 & 4 & 5 & 6 & 7 \\ 2 & 3 & 1 & 4 & 5 & 6 & 7 \end{pmatrix}, \alpha^{-1}\beta^{-1} = \begin{pmatrix} 1 & 2 & 3 & 4 & 5 & 6 & 7 \\ 1 & 2 & 3 & 4 & 7 & 5 & 6 \end{pmatrix}$$

Thus $\alpha\beta \neq \beta\alpha$, $(\alpha\beta)^1 \neq \beta^{-1}\alpha^{-1}$, $(\beta\alpha)^{-1} \neq \alpha^{-1}\beta^{-1}$.

(2) Let $\alpha, \beta \in S_x$ where $0(X) = 6$
and $X = \{a, b, c, d, i, j\}$ such that

$$\alpha = \begin{pmatrix} a & b & c & d & i & j \\ b & c & j & i & d & a \end{pmatrix} \text{ and}$$

$$\beta = \begin{pmatrix} a & b & c & d & i & j \\ a & c & i & j & b & d \end{pmatrix}$$

then
$$\alpha\beta^{-1} = \begin{pmatrix} a & b & c & d & i & j \\ b & c & j & i & d & a \end{pmatrix} \begin{pmatrix} a & c & i & j & b & d \\ a & b & c & d & i & j \end{pmatrix}$$

$$= \begin{pmatrix} a & b & c & d & i & j \\ i & b & d & c & j & a \end{pmatrix}$$

Also $(\alpha\beta^{-1})^6 = \begin{pmatrix} a & b & c & d & i & j \\ i & b & d & c & j & a \end{pmatrix}^6 = I$, and $(\alpha\beta^{-1})^m \neq I$ for $m \in Z^+$ and $m < 6$, thus $0(\alpha\beta^{-1}) = 6$

(3) If $\alpha \in S_6$ where $a = \begin{pmatrix} 1 & 2 & 3 & 4 & 5 & 6 \\ 2 & 3 & 6 & 5 & 4 & 1 \end{pmatrix}$

then since $\alpha^4 = I$, $0(\alpha) = 4$

as 4 is the least positive integer making $\alpha^4 = I$.

2.2 The group of permutations

In the preceeding sections we have seen that the set of all permutations S_x *over a finite set* X having 4 elements x, y, z, t satisfies all the group axioms for a non-abelian group under permutation product. Similarly if X is a finite set having any n elements then also elements of S_x satisfy all the group axioms for a non-abelian group (See problem 2.1). This group of $\lfloor n$ elements is called the group of permutations on n symbols where b.o. is the permutation multiplication and the group is a finite non-abelian group.

2.2.1. The group S_n

In its special case when n elements of set X are first n-natural numbers 1, 2, 3, ..., n (in place of any n elements) the group of permutation S_x is denoted by the notation S_n and is then called *the group of symmetries (or symmetric group)* of *degree n*.

This if $X = \{1, 2, 3, \ldots n\}$ then

$$S_n \left\{ \begin{pmatrix} 1 & 2 & 3 & \ldots & n \\ 1 & 2 & 3 & \ldots & n \end{pmatrix}, \begin{pmatrix} 1 & 2 & 3 & \ldots & n \\ 2 & 3 & 1 & \ldots & n \end{pmatrix}, \ldots, \right.$$
$$\left. \begin{pmatrix} 1 & 2 & 3 & \ldots & n \\ 1 & n & 2 & \ldots & 3 \end{pmatrix} \right\}$$

Where $0(S_n) = \lfloor n$.

Particular cases of the symmetric group S_n of degrees n arise in various significant symmetry considerations of a number of geometrical figures. We will study the group S_n in detail when $n = 3$ and $n = 4$.

2.2.2 The group S_3

When $n = 3$, the group S_3 of all permutations over the set of first 3 natural numbers 1, 2, 3 under operation of permutation multiplication has $\lfloor 3$ elements which are written in the following notations :

$$\rho_0 = \begin{pmatrix} 1 & 2 & 3 \\ 1 & 2 & 3 \end{pmatrix}, \rho_1 \begin{pmatrix} 1 & 2 & 3 \\ 2 & 3 & 1 \end{pmatrix}, \rho_2 = \begin{pmatrix} 1 & 2 & 3 \\ 3 & 1 & 2 \end{pmatrix}$$

and

$$\mu_1 = \begin{pmatrix} 1 & 2 & 3 \\ 1 & 3 & 2 \end{pmatrix}, \mu_2 = \begin{pmatrix} 1 & 2 & 3 \\ 3 & 2 & 1 \end{pmatrix}, \mu_3 = \begin{pmatrix} 1 & 2 & 3 \\ 2 & 1 & 3 \end{pmatrix}$$

The elements ρ_0, ρ_1, ρ_2 are obtained by moving anticlockwise and the elements μ_1, μ_2, μ_3 are obtained by moving clockwise along the three vertices of above triangle, starting from position 1.

Thus $S_3 = \{\rho_0, \rho_1, \rho_2, \mu_1, \mu_2, \mu_3\}$ where ρ_0 is the identity element.

Properties of elements of group S_3

(1) $\qquad \rho_0^2 = \rho_0, \; \rho_1^2 = \rho_2$ and $\rho_2^2 = \rho_1$

$\qquad \mu_1^2 = \rho_0, \; \mu_2^2 = \rho_0$ and $\mu_3^2 = \rho_0$

Thus, out of 6 elements of S_3 the four elements ρ_0, μ_1, μ_2 and μ_3 when squarred (under permutation product as the operation) give us the identity element, so that they satisfy the general quadratic equation $x^2 = e$, *i.e., the quadratic $x^2 = e$ has 4 solution if S_3.*

Also $\rho_0^2 = \rho_0 \Rightarrow \rho_0 = \rho_0^{-1}$; $\mu_1^2 = \rho_0 \Rightarrow \mu_1 = \mu_1^{-1}$, $\mu_2^2 = \rho_0 \Rightarrow \mu_2 = \mu_2^{-1}$, $\mu_3^2 = \rho_0 \Rightarrow \mu_3 = \mu_3^{-1}$. Thus the elements ρ_0, μ_1, μ_2 and μ_3 are their own inverses.

(2) $\rho_1 \rho_0 = \rho_1 = \rho_0 \rho_1$; $\rho_2 \rho_0 = \rho_2 = \rho_0 \rho_2$;

$\rho_1 \rho_2 = \rho_0 = \rho_2 \rho_1 \Rightarrow \rho_1 \rho_0 = \rho_0 \rho_1, \; \rho_2 \rho_0 = \rho_0 \rho_2$

and $\rho_1 \rho_2 = \rho_2 \rho_1$. Thus the element ρ_0, ρ_1, ρ_2 satisfy commutative law amongst themselves.

Also

$\mu_1 \mu_2 = \rho_2, \; \mu_2 \mu_1 = \rho_1, \; \mu_1 \mu_3 = \rho_1,$

$\mu_3 \mu_1 = \rho_2, \; \mu_2 \mu_3 = \rho_2, \; \mu_3 \mu_2 = \rho_1$

$\mu_2 \mu_3 \neq \mu_3 \mu_2$. Thus the elements μ_1, μ_2, μ_3 do note satisfy commutative law amongst themselves. Similarly $\mu_i \rho_j \neq \rho_j \mu_i$ for $i = 1, 2, 3; j = 1, 2$.

(3) The set S_3 with its above six elements forms a finite non-abelian group and has the following group table :

Permutation and Groups of Permutation

*	ρ_0	ρ_1	ρ_2	μ_1	μ_2	μ_3
	First Block			Second Block		
ρ_0	ρ_0	ρ_1	ρ_2	μ_1	μ_2	μ_3
ρ_1	ρ_1	ρ_2	ρ_0	μ_2	μ_4	μ_4
ρ_2	ρ_2	ρ_0	ρ_1	μ_3	μ_1	μ_2
	Third Block			Fourth Block		
μ_1	μ_1	μ_3	μ_2	ρ_0	ρ_2	ρ_1
μ_2	μ_2	μ_1	μ_3	ρ_1	ρ_0	ρ_2
μ_3	μ_3	μ_2	μ_1	ρ_2	ρ_1	ρ_0

Note : (1) If we divide the group-table of S_3 in 4 blocks, the elements in the first and second block move in cyclic (anticlockwise) order while in the third and fourth blocks, the lower most element in their first (and second) columns are raised on the top and rest two elements are copied to fill second (and third) columns of both the blocks.

(2) The first block of the table containing elements ρ_0, ρ_1 and ρ_2 is symmetrical along the principal diagonal and these elements form a separate (abelian) group of their own under permutation multiplication with ρ_0 as the identity element (it is easy to check all group axioms for them). The subset $\{\rho_0, \rho_1, \rho_2\}$ of S_3 is denoted by A_3 and is called the *Alternating group of* order three. Thus A_3 is a group within a group S_3. The groups within a group give rise to the important concept of *subgroups* of that group which has been taken up in the next chapter. The *group table of* A_3 is :

*	ρ_0	ρ_1	ρ_2
ρ_0	ρ_0	ρ_1	ρ_2
ρ_1	ρ_1	ρ_2	ρ_0
ρ_2	ρ_2	ρ_0	ρ_1

(* : Permutation multiplication)

2.2.3 The group S_4

When $n = 4$, the set S_n reduces to the set S_4, the set of all permutations over first 4 natural numbers 1, 2, 3, 4 which forms a (non-abelian) group under permutation multiplication and its has $\lfloor 4$ elements. Two of the subsets of S_4 which we denote respectively by D_4 and A_4 and which are particular cases of more general sets D_n and A_n have significant roles in group theory. We have discussed them in detail in 2.3.1 and 2.3.2 respectively.

2.3 Groups of Symmetries

In the Euclidean plane, consider a square whose centre O is at the origin and whose sides are parallel to the two axes. We can have *eight* different ways of moving the four vertices (individual points) of the square without disturbing the original position of the square. These movements are the *transformations* of the plane. Out of eight transformations, four are *rotations* and four are *reflections*. They are as follows :

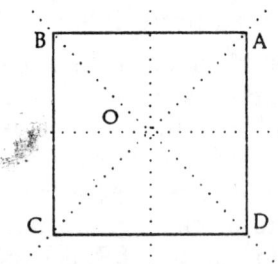

(i) The original position as given in the figure (rotation of the plane through O) is denoted by ρ_0.

(ii) Send A to B, B to C, C to D and D to A (rotation of the plane through $\pi/2$); denote it by v_1.

(iii) Send A to C, C to A, B to D and D to B (rotation through π); denote it by v_2.

(iv) Send A to D, B to A, C to B and D to C (rotation through $\frac{3\pi}{2}$); denote it by v_3.

(v) Send A to D, B to C, C to B and D to A (reflection in X-axis); denote it by τ_1.

Permutation and Groups of Permutation 95

(vi) Keep A and C unchanged, send B to D and D to B (reflection in the diagonal y = x); denote it by τ_2.

(vii) Send A to B, B to A, D to C and C to D (reflection in Y-axis); denote it by τ_3.

(viii) Keep B and D unchanged, send A to C and C to A (reflection in the diagonal y = – x); denote it by τ_4.

These transformations which leave the position of square unchanged are called the *symmetries of the square* and we have the following notations for them :

(a) Rotations

ρ_0 = rotation through O

v_1 = rotation through $\pi/2$

v_2 = rotation through π

v_3 = rotation through $3\pi/2$

(b) Reflections

τ_1 = reflection in the X-axis

τ_2 = reflection in the diagonal y = x

τ_3 = reflection in the Y-axis, and

τ_4 = reflection in the diagonal y = – x.

2.3.1 The group D_4

We will now see that these eight symmetries of square satisfy axioms of a non-abelian group :

(i) If f and g are any of the above eight symmetries then fg is one of them which is obtained by first applying g then f, similar to composition of any two functions which is defined as $(fg)(x) == f(g(x))$. For example, the combined effect of $v_1 \tau_2$ and the effect of τ_3 are the same, i.e., $v_1 \tau_2 = \tau_3$; the effects of $v_3 \tau_1$ and τ_4 are the same, i.e., $v_3 \tau_1 = \tau_4$, etc. Hence the combination of any two transformations obey the b.o. property.

(ii) Similarly combinations of these symmetries satisfy associative law as, for example, the effects of $v_1 (\tau_1 \tau_3)$ and $(v_1 \tau_1) \tau_3$ are the same. Infact

$$v_1 (t_1 \tau_3) = v_1 (v_2) = v_3, \text{ and}$$
$$(v_1 \tau_1) \tau_3 = (\tau_2) \tau_3 = v_3, \text{ i.e.,}$$
$$v_1 (\tau_1 \tau_3) = (v_1 \tau_1) \tau_3$$

This property can be seen for any three of the eight transformations.

(iii) The transformation ρ_0 acts like the identity element, for, it f is any symmetry of square the combined effect of $\rho_0 f$ satisfy

$$\rho_0 f = f \rho_0 = f$$

(iv) If f^2 denotes the repeated effect of a symmetry, we can easily see that

$$\rho_0^2 = \rho_0, \ v_2^2 = \rho_0, \ \tau_1^2 = \rho_0, \ \tau_2^2 = \rho_0, \ \tau_3^2 = \rho_0$$

$\tau_4^2 = \rho_0$ and also $v_1^2 = v_2 \ (\Rightarrow v_1^4 = \rho_0)$

$$v_3^2 = v_2 = (\Rightarrow v_3^4 = \rho_0)$$

Thus the symmetries ρ_0, v_2 and τ_1 to τ_4 are their own inverses while inv. of $v_1 = v_3$ and inv. of $v_3 = v_1$ so that each symmetry has its inverse from amongst the eight symmetries.

(v) The combined effects of $v_1 \tau_4$ and $\tau_4 v_1$ are not the same, as observe that $v_1 \tau_4 = \tau_1$ and $\tau_4 v_1 = \tau_3$, i.e., $v_1 \tau_4 \neq \tau_4 v_1$. Thus the combinations of symmetries of square do not satisfy commutative law.

From (i) to (v) above we conclude that the set $\{\rho_0, v_1, v_2, v_3, \tau_1, \tau_2, \tau_3, \tau_4\}$ of symmetries of square satisfies the axioms of a non-abelian group under combination of symmetries, i.e., under composition of two functions. This group is called the *group of symmetries of the square*.

If we rename the vertices A, B, C, D of square in the previous figure as 1, 2, 3, 4 respectively, we can describe the effect of each transformation, discussed in 2.3, in terms of the vertices as follows :

$$\rho_0 = \begin{pmatrix} 1 & 2 & 3 & 4 \\ 1 & 2 & 3 & 4 \end{pmatrix}$$

$$v_1 = \begin{pmatrix} 1 & 2 & 3 & 4 \\ 2 & 3 & 4 & 1 \end{pmatrix}$$

$$v_2 = \begin{pmatrix} 1 & 2 & 3 & 4 \\ 3 & 4 & 1 & 2 \end{pmatrix}$$

$$v_3 = \begin{pmatrix} 1 & 2 & 3 & 4 \\ 4 & 1 & 2 & 3 \end{pmatrix}; \text{ and}$$

$$\tau_1 = \begin{pmatrix} 1 & 2 & 3 & 4 \\ 4 & 3 & 2 & 1 \end{pmatrix}, \tau_2 = \begin{pmatrix} 1 & 2 & 3 & 4 \\ 1 & 4 & 3 & 2 \end{pmatrix}, \tau_3 = \begin{pmatrix} 1 & 2 & 3 & 4 \\ 2 & 1 & 4 & 3 \end{pmatrix}$$

$$\tau_4 = \begin{pmatrix} 1 & 2 & 3 & 4 \\ 3 & 2 & 1 & 4 \end{pmatrix}.$$

Thus symmetries of square are eight elements out of the $\lfloor 4$ elements of the group S_4. We denote the set of these eight elements by D_4. Thus

$D_4 = \{\rho_0, v_1, v_2, v_3, \tau_1, \tau_2, \tau_3, \tau_4\}$ and D_4 *is the group of symmetries of square.* By virtue of its having eight elements, D_4 is also called the *Octic group*. We have the following group-table for the group D_4 :

*	ρ_0	v_1	v_2	v_3	τ_1	τ_2	τ_3	τ_4
ρ_0	ρ_0	v_1	v_2	v_3	τ_1	τ_2	τ_3	τ_4
v_1	v_1	v_2	v_3	ρ_0	τ_2	τ_3	τ_4	τ_1
v_2	v_2	v_3	ρ_0	v_1	τ_3	τ_4	τ_1	τ_2
v_3	v_3	ρ_0	v_1	v_2	τ_4	τ_1	τ_2	τ_3
τ_1	τ_1	τ_4	τ_3	τ_2	ρ_0	v_3	v_2	v_1
τ_2	τ_2	τ_1	τ_4	τ_3	v_1	ρ_0	v_3	v_2
τ_3	τ_3	τ_2	τ_1	τ_4	v_2	v_1	ρ_0	v_3
τ_4	τ_4	τ_3	τ_2	τ_1	v_3	v_2	v_1	ρ_0

Group table of D_4

The subset D_n

In general, D_n ($n > 2$) is the subset having $2n$ elements of the group S_n and the elements of D_n describe the symmetries of a regular n-sided polygon and they form a non abelian group. The infinite family of group of symmetries D_n ($n = 3, 4, \ldots$) is called the family of *Dihedral groups*.

The case n = 2 : When $n = 2$, the symmetries that need to be considered will be a two-dimensional object (circle). In case of a circle, whose centre is at origin and whose radius is 1, the set of symmetries will have infinitely many rotations and reflections obtained by rotating and reflecting θ. These symmetries also satisfy axioms of a non-abelian group and this group is called the *orthogonal group* O_2. Thus O_2 is an infinite non-abelian group.

2.3.2 The group A_4

The other useful subset of the group S_4 is the set A_4 of 12 elements ($\frac{1}{2} O(S_4)$) and is given by

$A_4 = \{\rho_0, \nu_2, \tau_1, \tau_3, l, l_1, l_2, l_3, m_1, m_2, m_3, m_4\}$, where its elements stand for following permutations of the first-four natural numbers 1, 2, 3, 4 :

$$\rho_0 = \begin{pmatrix} 1 & 2 & 3 & 4 \\ 1 & 2 & 3 & 4 \end{pmatrix}, \qquad \nu_2 = \begin{pmatrix} 1 & 2 & 3 & 4 \\ 3 & 4 & 1 & 2 \end{pmatrix}$$

$$\tau_1 = \begin{pmatrix} 1 & 2 & 3 & 4 \\ 4 & 3 & 2 & 1 \end{pmatrix}, \qquad \tau_3 = \begin{pmatrix} 1 & 2 & 3 & 4 \\ 2 & 1 & 4 & 3 \end{pmatrix},$$

$$l = \begin{pmatrix} 1 & 2 & 3 & 4 \\ 2 & 3 & 1 & 4 \end{pmatrix}, \qquad l_1 = \tau_3 l = \begin{pmatrix} 1 & 2 & 3 & 4 \\ 3 & 2 & 4 & 1 \end{pmatrix},$$

$$l_2 = \nu_2 l = \begin{pmatrix} 1 & 2 & 3 & 4 \\ 1 & 4 & 2 & 3 \end{pmatrix}, \qquad l_3 = \tau_1 l = \begin{pmatrix} 1 & 2 & 3 & 4 \\ 4 & 1 & 3 & 2 \end{pmatrix};$$

$$m_1 = l^2 = \begin{pmatrix} 1 & 2 & 3 & 4 \\ 3 & 1 & 2 & 4 \end{pmatrix}, \qquad m_2 = \tau_3 l^2 = \begin{pmatrix} 1 & 2 & 3 & 4 \\ 1 & 3 & 4 & 2 \end{pmatrix}$$

$$m_3 = \nu_2 l^2 = \begin{pmatrix} 1 & 2 & 3 & 4 \\ 2 & 4 & 3 & 1 \end{pmatrix}, \qquad m_4 = \tau_1 l^2 = \begin{pmatrix} 1 & 2 & 3 & 4 \\ 4 & 2 & 1 & 3 \end{pmatrix}$$

The subset A_4 forms a group of its own under permutation multiplication and it is one from the family of *Alternating groups*. From the group table of A_4, given on page 113, it is evident that the four elements ρ_0, ν_2, τ_1 and τ_3 of A_4 form Klein's four group K_4.

*	ρ_0	τ_3	v_2	τ_1	l	l_1	l_2	l_3	m_1	m_2	m_3	m_4
ρ_0	ρ_0	τ_3	v_2	τ_1	l	l_1	l_2	l_3	m_1	m_2	m_3	m_4
τ_3	τ_3	ρ_0	τ_1	v_2	l_1	l	l_3	l_2	m_2	m_1	m_4	m_3
v_2	v_2	τ_1	ρ_0	τ_3	l_2	l_3	l	l_1	m_3	m_4	m_1	m_2
τ_1	τ_1	v_2	τ_3	ρ_0	l_3	l_2	l_1	l	m_4	m_3	m_2	m_1
l	l	l_2	l_3	l_1	m_1	m_3	m_4	m_2	ρ_0	v_2	τ_1	τ_3
l_1	l_1	l_3	l_2	l	m_2	m_4	m_3	m_1	τ_3	τ_1	v_2	ρ_0
l_2	l_2	l	l_1	l_3	m_3	m_1	m_2	m_4	v_2	ρ_0	ρ_3	τ_2
l_3	l_3	l_1	l	l_2	m_4	m_2	m_1	m_3	τ_1	τ_3	ρ_0	v_2
m_1	m_1	m_4	m_2	m_3	ρ_0	τ_1	τ_3	v_2	l	l_3	l_1	l_2
m_2	m_2	m_3	m_1	m_4	τ_3	v_2	ρ_0	τ_1	l_1	l_2	l	l_3
m_3	m_3	m_2	m_4	m_1	v_2	τ_3	τ_1	ρ_0	l_2	l_1	l_3	l
m_4	m_4	m_1	m_3	m_2	τ_1	ρ_0	v_2	τ_3	l_3	l	l_2	l_1

Group table A_4

2.3.3 Other groups of Symmetry

We now consider the groups of symmetries of a triangle and a rectangle ($n = 3, n = 4$) :

(1) *Group of symmetries of an isoceles triangle* : The triangle is symmetrical about the line AM, the perpendicular bisector. Here symmetries are two :

(*i*) rotation through O, and (*ii*) reflection in AM.

These two symmetries, in terms of vertices 1, 2, 3 of the triangle, are :

Figure

ρ_0 (rotation through O)

$$= \begin{pmatrix} 1 & 2 & 3 \\ 1 & 2 & 3 \end{pmatrix}$$

μ_1 (reflection in AM)

$$= \begin{pmatrix} 1 & 2 & 3 \\ 1 & 3 & 2 \end{pmatrix}$$

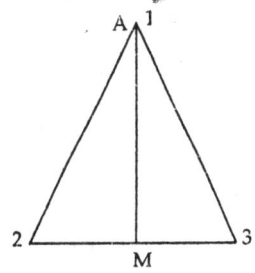

and the group $\{\rho_0, \mu_1\}$ is the group of symmetries of an isoceles triangle. Notice that $\{\rho_0, \mu_1\}$ is a subset of the group S_3.

(2) *Group of symmetries of an equilateral triangle*: In this case we have six symmetries – three rotations and three reflections. They are:

(i) rotation through $\frac{2\pi}{3}$

(ii) rotation through $\frac{4\pi}{3}$

(iii) rotation through $\frac{6\pi}{3}$
(or through 2π), and

(iv) reflection in AM

(v) reflection in BL

(vi) reflection in CN

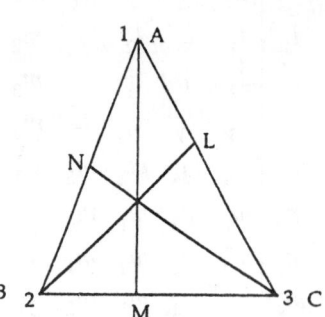

By renaming the vertices of the triangle as 1, 2 3, as shown, we see that above six symmetries in terms of their notations are given by:

ρ_0 (rotation through 2π) = $\begin{pmatrix} 1 & 2 & 3 \\ 1 & 2 & 3 \end{pmatrix}$;

r_1 (rotation through $\frac{2\pi}{3}$) = $\begin{pmatrix} 1 & 2 & 3 \\ 2 & 3 & 1 \end{pmatrix}$;

ρ_2 (rotation through $\frac{4\pi}{3}$) = $\begin{pmatrix} 1 & 2 & 3 \\ 3 & 2 & 1 \end{pmatrix}$; and

μ_1 (reflection in AM) = $\begin{pmatrix} 1 & 2 & 3 \\ 1 & 3 & 2 \end{pmatrix}$;

μ_2 (reflection in BL) = $\begin{pmatrix} 1 & 2 & 3 \\ 3 & 2 & 1 \end{pmatrix}$;

μ_3 (reflection in CN) = $\begin{pmatrix} 1 & 2 & 3 \\ 2 & 1 & 3 \end{pmatrix}$.

Permutation and Groups of Permutation

These six elements form a group of their own and this group is one from the family of Dihedral groups D_n when $n = 3$ having $2n$ i.e., 6 elements. This

$D_3 = \{P_0, P_1, P_2, \mu_1, \mu_2, \mu_3\}$ and is called the *group of symmetries of equilateral triangle*. Notice that the elements of D_3 are the same as those of group S_3, the group of all permutations over first three natural numbers 1, 2, 3. Hence S_3 is also called the group D_3 of symmetries of an equilateral triangle.

(*iii*) *Group of symmetries of a rectangle* : Let GM and LN be the lines joining the mid points of opposite sides of rectangle ABCD, as shown. Then the symmetries are :

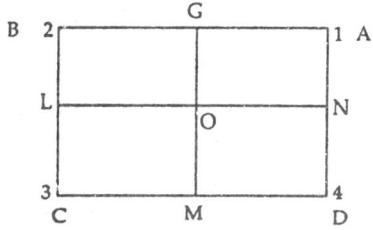

(*i*) rotation through O $= \begin{pmatrix} 1 & 2 & 3 & 4 \\ 1 & 2 & 3 & 4 \end{pmatrix}$;

(*ii*) rotation about O

through $\pi = \begin{pmatrix} 1 & 2 & 3 & 4 \\ 3 & 4 & 1 & 2 \end{pmatrix}$;

(*iii*) reflection in LN $= \begin{pmatrix} 1 & 2 & 3 & 4 \\ 4 & 3 & 2 & 1 \end{pmatrix}$,

(*iv*) reflection in GM $= \begin{pmatrix} 1 & 2 & 3 & 4 \\ 2 & 1 & 4 & 3 \end{pmatrix}$,

which when denoted in terms of notations of the group of symmetries of a square are respectively ρ_0, v_2, τ_1 and τ_3. These elements form a group of their own under permutation multiplication and the group $\{r_0, v_2, \tau_1, \tau_3\}$ is the *group of symmetries of a rectangle*. From the group table of D_4, it follows that this groups is Klein's group K_4.

In the preceeding sections, we have seen that symmetries of certain geometrical figures can easily be described through group of permutations over first n-natural numbers. In fact, it is very convenient to study symmetries through groups. Recently the symmetries in nature including those in molecules, crystals and elementary particles have been studied via group theory. The interested reader may refer to H. Weyl, S. Sternberg and other references given in the end for further details.

2.4 Cyclic Permutations and Cycles

For some permutations it is possible to write them in a *single-row notation*, in place of two-row notation. Consider the following permutations form S_6 and S_5 respectively :

$$\alpha = \begin{pmatrix} 1 & 2 & 3 & 4 & 5 & 6 \\ 2 & 3 & 4 & 5 & 6 & 1 \end{pmatrix}, \quad \beta = \begin{pmatrix} 1 & 2 & 3 & 4 & 5 & 6 \\ 2 & 4 & 1 & 3 & 5 & 6 \end{pmatrix}$$

and $\gamma = \begin{pmatrix} 1 & 2 & 3 & 4 & 5 \\ 3 & 2 & 5 & 1 & 4 \end{pmatrix}$

These permutations can be put in a single row as follows :

$\alpha = (1\ 2\ 3\ 4\ 5\ 6)$, $\beta = (1\ 2\ 4\ 3)$, $\gamma = (1\ 3\ 5\ 4)$.

The elements put in single row satisfy following :

(i) The image of an element in the single row is the one which follows it and the image of last element is the first elements of the row, *i.e.*, image of rth element in the row = $(r + 1)^{th}$ element and, image of last element in the row = the first element.

(ii) If an element of the permutation given in two-row notation is not mentioned in the one-row notation, it's image will be the element itself. For example, there is no mention of 2 in the one-row notation of permutation γ so $2\gamma = 2$, *i.e.*, 2 will be carried over to 2, itself, as given in two-row notation for γ. Similarly $5\beta = 5$, as there is no mention of 5 in the single row of β.

Definition 2.4.1 : If it is possible to write a given permutation in one-row notation (following above two points), then the permutation is called a *cyclic permutation* and the single row is called *cycle* of the permutation.

Definition 2.4.2 : The number of elements used in a cycle is called the *length* of the cycle and it is ≥ 2.

Thus α, β, γ given above are cyclic permutations and (1 2 3 4 5 6), (1 2 4 3), (1 3 5 4) are, respectively, their cycles; length of their cycles being 6, 4 and 2.

Examples

(1) The elements $\rho_1 = \begin{pmatrix} 1 & 2 & 3 \\ 2 & 3 & 1 \end{pmatrix}$, $\rho_2 = \begin{pmatrix} 1 & 2 & 3 \\ 3 & 1 & 2 \end{pmatrix}$,

$\mu_1 = \begin{pmatrix} 1 & 2 & 3 \\ 1 & 3 & 2 \end{pmatrix}$, $\mu_2 = \begin{pmatrix} 1 & 2 & 3 \\ 3 & 2 & 1 \end{pmatrix}$ and $\mu_3 = \begin{pmatrix} 1 & 2 & 3 \\ 2 & 1 & 3 \end{pmatrix}$

of the groups S_3 are all cyclic, as

$\rho_1 = (1\ 2\ 3)$, $\rho_2 = (1\ 3\ 2)$, $\mu_1 = (2\ 3)$,

$\mu_2 = (1\ 3)$ and $\mu_3 = (1\ 2)$.

(2) The elements $\tau_1 = \begin{pmatrix} 1 & 2 & 3 & 4 \\ 4 & 3 & 2 & 1 \end{pmatrix}$, $v_2 = \begin{pmatrix} 1 & 2 & 3 & 4 \\ 3 & 4 & 1 & 2 \end{pmatrix}$ of D_4 are

not cyclic permutations as we can not write them in one-row notation.

2.4.1 Multiplication of cycles

We can multiply two or more cycles from the same set of permutations to get another cycle from that set. The multiplication of cycles is done by first converting them into their permutations and then performing the rule of permutation multiplication. The newly obtained permutation can then be converted back to its cycle. Here are two examples of cycle multiplication :

Let

$\sigma = (1\ 2\ 3)$ and $\delta = (5\ 6\ 4\ 1)$ be two cycles in S_6. Converting them back to their permutations and then taking the product, we get

$$\sigma\delta = \begin{pmatrix} 1 & 2 & 3 & 4 & 5 & 6 \\ 2 & 3 & 1 & 4 & 5 & 6 \end{pmatrix} \begin{pmatrix} 1 & 2 & 3 & 4 & 5 & 6 \\ 5 & 2 & 3 & 1 & 6 & 4 \end{pmatrix}$$

$$\sigma\delta = \begin{pmatrix} 1 & 2 & 3 & 4 & 5 & 6 \\ 2 & 3 & 5 & 1 & 6 & 4 \end{pmatrix} = (1\ 2\ 3\ 5\ 6\ 4)$$

Next, $\quad \delta\sigma = \begin{pmatrix} 1 & 2 & 3 & 4 & 5 & 6 \\ 5 & 3 & 1 & 2 & 6 & 4 \end{pmatrix} = (1\ 5\ 6\ 4\ 2\ 3)$

Thus :

(i) given permutations σ and δ are cyclic and their products $\sigma\delta$ and $\delta\sigma$ are also cyclic.

(ii) $\sigma\delta \neq \delta\sigma$ (notice that element 1 is common between the two cycles)

Definition 2.4.3 : Two cycles from a set of permutations are said to be *disjoint* if they have no element common in them.

The cycles σ and δ from S_6, given above, are not disjoint as $\sigma \cap \delta = \{1\}$. The cycles $p = (1\ 3\ 5)($ and $q = (2\ 6\ 8\ 9)$ from S_9 are disjoint as nothing is common between them.

2.4.2 Product of cycles and commutative law

We noticed in above example of cycle multiplication that $\sigma\delta \neq \delta\sigma$ and σ and δ are not disjoint. We will now see that the product of disjoint cycles from a set of permutations obey commutative law.

If $p, q \in S_9$ and they are as given above then

$pq = (1\ 3\ 5)\ (2\ 6\ 8\ 9)$

$= \begin{pmatrix} 1 & 2 & 3 & 4 & 5 & 6 & 7 & 8 & 9 \\ 3 & 2 & 5 & 4 & 1 & 6 & 7 & 8 & 9 \end{pmatrix} \begin{pmatrix} 1 & 2 & 3 & 4 & 5 & 6 & 7 & 8 & 9 \\ 1 & 6 & 3 & 4 & 5 & 8 & 7 & 9 & 2 \end{pmatrix}$

$= \begin{pmatrix} 1 & 2 & 3 & 4 & 5 & 6 & 7 & 8 & 9 \\ 3 & 6 & 5 & 4 & 1 & 8 & 7 & 9 & 2 \end{pmatrix}$

and, $\quad qp = (2\ 6\ 8\ 9)\ (1\ 3\ 5) = \begin{pmatrix} 1 & 2 & 3 & 4 & 5 & 6 & 7 & 8 & 9 \\ 3 & 6 & 5 & 4 & 1 & 8 & 7 & 9 & 2 \end{pmatrix} = pq$

Thus when cycles from a set are disjoint their product obeys C.L. Check it for yourself in the case of cycles $u = (2\ 4\ 6)$ and $v = (1\ 3\ 5\ 7)$ from the set S_7; you will get $uv = vu = \begin{pmatrix} 1 & 2 & 3 & 4 & 5 & 6 & 7 \\ 3 & 4 & 5 & 6 & 7 & 2 & 1 \end{pmatrix}$.

2.4.3 A permutation and its disjoint cycles

We will now see that every permutation can be *uniquely* decomposed (expressed) as a product of disjoint cycles.

For examples, if $\alpha \in S_6$ and

$$\alpha = \begin{pmatrix} 1 & 2 & 3 & 4 & 5 & 6 \\ 6 & 5 & 2 & 4 & 3 & 1 \end{pmatrix}$$ then we can decompose α as,

$$\alpha = \begin{pmatrix} 1 & 2 & 3 & 4 & 5 & 6 \\ 6 & 5 & 2 & 4 & 3 & 1 \end{pmatrix} = (1\ 6)\ (2\ 5\ 3)$$

so that α is expressed as a product of two disjoint cycles. For doing so, we proceed as follows :

(i) Start with 1. Since $1\alpha = 6$ and $6\alpha = 1$, i.e., $1 \to 6$ and $6 \to 1$, so the first cycle in the right side product is (1, 6).

(ii) Again, start with the next element 2. Since $2\alpha = 5$, $5\alpha = 3$, $3\alpha = 2$, the second cycle in the product is (2 5 3).

(iii) The elements having the same image are not put in any of the cycles. Here the element 4 has been left out in the product as $4\alpha = 4$. Thus given permutation α = (16) (2 5 3).

(iv) By multiplying the cycles (16) and (2 5 3), as described in 2.4.1, we get back the permutation α. Also, the product is unique.

Similarly if $\beta \in S_9$ and

$$\beta = \begin{pmatrix} 1 & 2 & 3 & 4 & 5 & 6 & 7 & 8 & 9 \\ 2 & 3 & 1 & 4 & 8 & 6 & 9 & 7 & 5 \end{pmatrix}$$

then
$\beta = (1\ 2\ 3)\ (5\ 8\ 7\ 9);$

if $\gamma \in S_{10}$ and

$$\gamma = \begin{pmatrix} 1 & 2 & 3 & 4 & 5 & 6 & 7 & 8 & 9 & 10 \\ 1 & 3 & 5 & 7 & 9 & 8 & 4 & 10 & 2 & 6 \end{pmatrix}$$

then
$\gamma = (2\ 3\ 5\ 9)\ (4\ 7)\ (6\ 8\ 10).$

Likewise you can express $\delta \in S_{10}$, where

$$\delta = \begin{pmatrix} 1 & 2 & 3 & 4 & 5 & 6 & 7 & 8 & 9 & 10 \\ 2 & 9 & 7 & 1 & 5 & 10 & 8 & 3 & 4 & 6 \end{pmatrix}.$$

Order of a Permutation and Cycles : We have just seen that any permutation can be uniquely expressed as a product of disjoint cycles. Such decomposition of a permutation is useful in finding its *order*. The order of a permutation which we defined in 2.1.3 is also equal to the LCM of lengths of *disjoint cycles* in the product of a permutation.

Examples

From above, we have

$\alpha = (1\ 6)\ (2\ 5\ 3)$

$\beta = (1\ 2\ 3)\ (5\ 8\ 7\ 9)$, and

$\gamma = (2\ 3\ 5\ 9)\ (4\ 7)\ (6\ 8\ 10)$

so that

$O(\alpha) = $ LCM $\{2, 3\} = 6$ (You can verify that $\alpha^6 = I$ and 6 is least such positive integer),

$O(\beta) = $ LCM $\{3, 4\} = 12$

$O(\gamma) = $ LCM $\{4, 2, 3\} = 12$.

Note : If cycles in a product are *not disjoint* then the order of permutation is *not equal* to the LCM of lengths of cycles (see solved problem 2.6).

Definition 2.4.2 : A cycle of length 2 is called a *transposition*.

For example (1 3) is a transposition in S_3. The permutation corresponding to the transposition (1 3) in S_3 is $\begin{pmatrix} 1 & 2 & 3 \\ 3 & 2 & 1 \end{pmatrix}$.

2.4.4 A permutation and its transpositions

Every permutation can also be decomposed as product of a number of transpositions. But this decomposition is *not unique*. For example if $k \in S_5$

and

$$k = \begin{pmatrix} 1 & 2 & 3 & 4 & 5 \\ 1 & 3 & 5 & 2 & 4 \end{pmatrix}$$ then we can decompose

k as,

$$k = (2\ 3\ 5\ 4) = (2\ 3)\ (2\ 5)\ (2\ 4) \qquad \ldots(1)$$

so that permutation k has been expressed as product of 3 transpositions. For doing so we proceed as follows :

(*i*) Convert the permutation into product of its cycle(s).

(*ii*) Then each cycle is decomposed as product of transposition by keeping the first element (2 in k) fixed and varying the second element in each transposition. Thus

$(2\ 3\ 5\ 4) = (2\ 3)\ (2\ 5)\ (2\ 4)$

(*iii*) By multiplying the transpositions (2 3), (2 5) and (2 4), by the method of cycle multiplication as described in 2.4.1, we get back the permutation k. Further, since $(2\ 3\ 5\ 4) = (3\ 5\ 4\ 2)$ (a cycle does not change till its cyclic order is maintained) thus $k = (3\ 5\ 4\ 2) = (3\ 5)\ (3\ 4)\ (3\ 2)$ by above method. Thus the decomposition (1) is not unique.

Similarly if $j \in S_7$ and

$$j = \begin{pmatrix} 1 & 2 & 3 & 4 & 5 & 6 & 7 \\ 2 & 1 & 4 & 7 & 3 & 6 & 5 \end{pmatrix} \text{ then } j = (1\ 2)\ (3\ 4\ 7\ 5)$$

$= (1\ 2)\ \{(3\ 4)\ (3\ 7)\ (3\ 5)\}$

so the j is expressible as product of 4 transpositions. We can also write $j = (1\ 2)\ (4\ 7\ 5\ 3) = (1\ 2)\ \{(4\ 7)\ (4\ 5)\ (4\ 3)\}$ or as $j = (1\ 2)\ (7\ 5\ 3\ 4) = (1\ 2)\ \{(7\ 5)\ (7\ 3)\ (7\ 4)\}$

Even and Odd permutations : The number of transpositions in decomposition of a permutation into product of cycles decides an important characteristic of the permutation, which we define now

Definition 2.4.3 : A permutation is said to be an *even permutation* if it can be expressed as product of an even number of transpositions. Otherwise it is said to be an *odd permutation*.

Examples

From above, $k = (2\ 3)(2\ 5)(2\ 4)$, which is product of three (odd) transpositions so that k is an odd permutation. Also, $j = (1\ 2)(3\ 4)(3\ 7)(3\ 5)$, which is product of four (even) transpositions so j is an even permutation.

Note : The distribution of even and odd permutations in a set of all permutations is 50-50. Since a cycle of length r can be expressed as product of $(r - 1)$ transpositions (see 2.4.3), a cycle of length r will represent an even permutation if r is odd number and a cycle of length r will represent an odd permutation if r is even number. Thus every transposition is an odd permutation.

2.4.5 Illustrative Examples

(1) The permutations $p \in S_8$ and $q \in S_9$, where

$$p = \begin{pmatrix} 1 & 2 & 3 & 4 & 5 & 6 & 7 & 8 \\ 4 & 5 & 3 & 6 & 1 & 2 & 8 & 7 \end{pmatrix} \text{ and}$$

$$q = \begin{pmatrix} 1 & 2 & 3 & 4 & 5 & 6 & 7 & 8 & 9 \\ 3 & 6 & 4 & 2 & 5 & 1 & 7 & 8 & 9 \end{pmatrix}$$

can be decomposed respectively as

$$p = (1\ 4\ 6\ 2\ 5)(7\ 8) = \{(1\ 4)(1\ 6)(1\ 2)(1\ 5)\}(7\ 8)$$

so that p is an odd permutation in set S_8
and

$$q = (1\ 3\ 4\ 2\ 6) = (1\ 3)(1\ 4)(1\ 2)(1\ 6),$$

so that q is an even permutation in set S_9.

(2) *The identity permutation is an even permutation* as it can be expressed as product of even number of transpositions.

For example, $I \in S_x$ where
$X = \{a_1, a_2, \ldots a_n\}$ then $I = \begin{pmatrix} a_1 & a_2 & a_n \\ a_1 & a_2 & a_n \end{pmatrix}$
$= (a_1\ a_2)(a_2\ a_1)$

Similarly $\rho_0 \in S_3$ where

$$\rho_0 = \begin{pmatrix} 1 & 2 & 3 \\ 1 & 2 & 3 \end{pmatrix}$$

$$= (1\ 2)\ (2\ 1)$$

(3) Out of total six elements of S_6, the elements ρ_0, ρ_1, ρ_2 are even and the elements μ_1, μ_2, μ_3 are odd permutations, as

ρ_0 = identity permutation

$$\rho_1 = \begin{pmatrix} 1 & 2 & 3 \\ 2 & 3 & 1 \end{pmatrix} = (1\ 2\ 3) = (1\ 2)\ (1\ 3)$$

(product of two transpositions)

$$\rho_2 = \begin{pmatrix} 1 & 2 & 3 \\ 3 & 1 & 2 \end{pmatrix} = (1\ 3\ 2) = (1\ 3)\ (1\ 2);\ \text{and}$$

$$\mu_1 = \begin{pmatrix} 1 & 2 & 3 \\ 1 & 3 & 2 \end{pmatrix} = (2\ 3)\ \text{(one transposition)}$$

$$\mu_2 \begin{pmatrix} 1 & 2 & 3 \\ 3 & 2 & 1 \end{pmatrix} = (1\ 3)$$

$$\mu_3 = \begin{pmatrix} 1 & 2 & 3 \\ 2 & 1 & 3 \end{pmatrix} = (1\ 2).$$

Recall from 2.2.2 that we have denoted the set $\{\rho_0, \rho_1, \rho_2\}$ by A_3

Thus in $S_3 = \{\rho_0, \rho_1, \rho_2, \mu_1, \mu_2, \mu_3\}$,

$A_3 = \{\rho_0, \rho_1, \rho_2\}$ which is the set of all even permutations of S_3,

and

$S_3 - A_3 = \{\mu_1, \mu_2, \mu_3\}$, which is the set of all odd permutations of S_3.

2.5 The set A_n and A_n-groups

We will see in Theorem 2.6.2 that out of $\lfloor n$ elements of the set S_x of all permutations over n elements, 50% are even permutations and the remaining 50% are odd. In particular the set of even permutations of S_n, set of all permutations over first n natural numbers 1, 2, ...n, is denoted by A_n. Thus if $\alpha \in S_n$ and α is even permutation then $\alpha \in A_n$. Thus $A_n \subseteq S_n$ and $O(A_n) = \frac{1}{2} \lfloor n$. The set of odd permutations of S_n is then $S_n - A_n$ and $O(S_n - A_n) = \frac{1}{2} \lfloor n$.

As we described in example 3 above the set of even permutations of set S_3 is $A_3 = \{\rho_0, \rho_1, \rho_2\}$. Similarly the set of A_4 of 12 elements as given in 2.3.2 is the set of all even permutations of set S_4. In example 1 above, since $q = (1\ 3\ 4\ 2\ 6)$ is an even permutation of S_9, therefore $q \in A_9$. In the same example $p \notin A_8$ as p is an odd permutation.

In general, the set A_n ($n \geq 2$) of all even permutations of group S_n forms a group of its own under operation of permutation multiplication and the group A_n is called *Alternating group*. (see solved problem 2.6). We have already seen the examples of alternating groups A_3 and A_4 in the previous sections.

2.6 General properties of permutations

Theorem 2.6.1 : If a permutation α is a expressed as a product of m_1 transpositions as well as a product of m_2 transpositions then either both m_1 and m_2 are even or both are odd numbers.

Proof : Consider the product

$$P = \prod_{i<j} (x_i - x_j), i, j = 1, 2, \ldots n. \text{ Then}$$

$$\begin{aligned}P = &(x_1 - x_2)(x_1 - x_3) \ldots (x_1 - x_n). \\ &(x_2 - x_3)(x_2 - x_4) \ldots (x_2 - x_n). \\ &\ldots \ldots (x_{n-2} - x_{n-1})(x_{n-2} - x_n) \\ &\ldots \ldots \ldots (x_{n-1} - x_n) \end{aligned} \quad \ldots(1)$$

By applying a transposition $t = (x_i\ x_j)$, $i < j$, (t means to interchange x_i and x_j) on the product P, we see that

$tP = (x_i\ x_j) P = -P$, because

Permutation and Groups of Permutation

(i) factors of P in (1) not having x_i and x_j are unaffected by the transposition,

(ii) the factor having x_i and x_j both will change its sign and hence that of P, and

(iii) the factors of P having either x_i or x_j but not both are unaffected by this transposition.

Similarly if two transpositions t *and* u are applied on P, we have

$$(t\ u)\ P = (-1)^2\ P, \text{ and}$$

if a permutation α contains product of m_1 transpositions then

$\alpha P = (-1)^{m_1} P$, or if α contains product of m_2 transpositions then

$$\alpha P = (-1)^{m_2} P \text{ so that}$$

$$(-1)^{m_1} P = (-1)^{m_2} P \Rightarrow P = (-1)^{m_1 - m_2} P$$

$\Rightarrow m_1, m_2$ are either both even or both odd. Examples given in 2.4.4. illustrate this theorem.

Theorem 2.6.2 : If X is a set of n elements then out of $\lfloor n$ elements of the set Sx, half are even and half re odd permutations.

Proof : Out of $\lfloor n$ permutations in Sx, let $E = \{E_1, E_2, ... E_i\}$ be set of i distinct even permutations and $O = \{O_1, O_2, ... O_j\}$ be set of j distinct odd permutations, so that

$$i + j = \lfloor n \qquad \qquad ...(1)$$

Let t be a transposition then

$t\ E = \{t\ E_1, t\ E_2, ..., tE_i\}$, where each of $tE_1, tE_2, ..., tE_i$ are distinct odd permutation, (they are distinct, for , if $tE_r = tE_s$ then $E_r \neq E_s$ by LCL in S_x), so $tE_1, tE_2, ..., tE_i \in O$

$$\Rightarrow i \leq j \qquad \qquad ...(2)$$

Similarly in $t\ O = \{tO_1, tO_2, ... tO_j\}$, each of $iO_1, tO_2, ..., tO_j$ are distinct even permutations and hence they belong to the set E. Thus $j \leq i$ ...(3). By (2) and (3), $i = j$ which by (1) implies that $O(E) = \frac{1}{2} \lfloor n$ and $O(O) = \frac{1}{2} \lfloor n$.

Solved Problems Set

Problem 2.1 : Let $X = \{a_1, a_2, a_3, \ldots a_n\}$ be a the set of n elements then show that the set S_x of all permutations over X forms a non-abelian finite group under the operation of permutation multiplication.

Solution : Here

$$S_x = \left\{ \begin{pmatrix} a_1 \ a_2 \ a_3 \ \ldots \ a_n \\ a_2 \ a_1 \ a_n \ \ldots \ a_3 \end{pmatrix}, \begin{pmatrix} a_1 \ a_2 \ a_3 \ \ldots \ a_n \\ a_n \ a_1 \ a_3 \ \ldots \ a_2 \end{pmatrix}, \right.$$

$$\left. \ldots \begin{pmatrix} a_1 \ a_2 \ a_3 \ \ldots \ a_n \\ a_1 \ a_n \ a_3 \ \ldots \ a_2 \end{pmatrix} \right\}$$

and $O(S_x) = \lfloor n$. Let α, β, γ denote respectively the permutations given in S_x.

(i) We know that permutation product is a b.o.

For example, here $\alpha\beta = \begin{pmatrix} a_1 \ a_2 \ a_3 \ \ldots \ a_n \\ a_1 \ a_n \ a_2 \ \ldots \ a_3 \end{pmatrix} \in S_x$.

(ii) $\alpha\,(\beta\gamma) = \begin{pmatrix} a_1 \ a_2 \ a_3 \ \ldots \ a_n \\ a_2 \ a_1 \ a_n \ \ldots \ a_3 \end{pmatrix} \left\{ \begin{pmatrix} a_1 \ a_2 \ a_3 \ \ldots \ a_n \\ a_2 \ a_1 \ a_3 \ \ldots \ a_n \end{pmatrix} \right\}$

$$= \begin{pmatrix} a_1 \ a_2 \ a_3 \ \ldots \ a_n \\ a_1 \ a_2 \ a_n \ \ldots \ a_3 \end{pmatrix}$$

Similarly $(\alpha\beta)\,\gamma = \begin{pmatrix} a_1 \ a_2 \ a_3 \ \ldots \ a_n \\ a_1 \ a_2 \ a_n \ \ldots \ a_3 \end{pmatrix} = \alpha\,(\beta\gamma)$

Thus elements of S_x satisfy associative law.

(iii) If $I = \begin{pmatrix} a_1 \ a_2 \ a_3 \ \ldots \ a_n \\ a_1 \ a_2 \ a_3 \ \ldots \ a_n \end{pmatrix}$

then $I \in S_x$ and $\alpha I = I\alpha = \alpha$ for any $\alpha \in S_x$, thus I is the identity elements in S_x

Permutation and Groups of Permutation

(*iv*) By definition of inverse

$$\alpha^{-1} = \begin{pmatrix} a_2 \, a_1 \, a_n \, \ldots \, a_3 \\ a_1 \, a_2 \, a_3 \, \ldots \, a_n \end{pmatrix}$$

then

$$\alpha\alpha^{-1} = \begin{pmatrix} a_1 \, a_2 \, a_3 \, \ldots \, a_n \\ a_2 \, a_1 \, a_n \, \ldots \, a_3 \end{pmatrix} \begin{pmatrix} a_2 \, a_1 \, a_n \, \ldots \, a_3 \\ a_1 \, a_2 \, a_3 \, \ldots \, a_n \end{pmatrix}$$

$$= \begin{pmatrix} a_1 \, a_2 \, a_3 \, \ldots \, a_n \\ a_1 \, a_2 \, a_3 \, \ldots \, a_n \end{pmatrix} = I;$$

similarly $\alpha^{-1} \alpha = I$. Thus each element in S_x has its inverse in S_x.

(*v*) Finally $\quad \alpha\beta = \begin{pmatrix} a_1 \, a_2 \, a_3 \, \ldots \, a_n \\ a_2 \, a_1 \, a_n \, \ldots \, a_3 \end{pmatrix} \begin{pmatrix} a_1 \, a_2 \, a_3 \, \ldots \, a_n \\ a_n \, a_1 \, a_3 \, \ldots \, a_2 \end{pmatrix}$

$$= \begin{pmatrix} a_1 \, a_2 \, a_3 \, \ldots \, a_n \\ a_1 \, a_n \, a_2 \, \ldots \, a_3 \end{pmatrix}$$

$$\beta\alpha = \begin{pmatrix} a_1 \, a_2 \, a_3 \, \ldots \, a_n \\ a_3 \, a_2 \, a_n \, \ldots \, a_1 \end{pmatrix}$$

Thus S_x is a non-abelian group of finite order.

Problem 2.2 : (*i*) Show that

$(x_1 \, x_2 \, \ldots \, x_n)^{-1} = (x_n \, x_{n-1} \, x_{n-2} \, \ldots \, x_3 \, x_2 \, x_1)$

(*ii*) Give examples from S_n the if f, g, h are any 3 cycles then

$(fg)^{-1} = g^{-1} \, f^{-1}$ and $(fgh)^{-1} = h^{-1} \, g^{-1} \, f^{-1}$

Solution : (*i*) Since

$$(x_1 \, x_2 \, x_3 \, \ldots \, x_n) \, (x_n \, x_{n-1} \, x_{n-2} \, \ldots \, x_3 \, x_2 \, x_1)$$

$$= \begin{pmatrix} x_1 \, x_2 \, x_3 \, \ldots \, x_n \\ x_2 \, x_3 \, \ldots \, \ldots \, x_1 \end{pmatrix} \begin{pmatrix} x_1 \, x_2 \, x_3 \, \ldots \, x_{n-1} \, x_n \\ x_n \, x_1 \, x_2 \, \ldots \, x_{n-2} \, x_{n-1} \end{pmatrix}$$

$$= \begin{pmatrix} x_1 \, x_2 \, x_3 \, \ldots \, x_n \\ x_1 \, x_2 \, \ldots \, x_n \end{pmatrix} = I$$

$\Rightarrow (x_1 \, x_2 x_3 \, \ldots \, x_n)^{-1} = (x_n \, x_{n-1} \, \ldots \, x_3 \, x_2 \, x_1)$

(ii) Let $f = (1\ 2\ 3\ \ldots\ n)$,
$g = (1\ 3\ 2\ \ldots\ n)$ and $h = (1\ n\ 3\ \ldots\ 2)$
and follow (i).

Problem 2.3 : Show that

(i) the product of two even permutations is an even permutation,

(ii) the product of two odd permutations is an even permutation

(iii) the product of an even (odd) permutation and an odd (even) permutation is an odd permutation

(iv) the inverse of an even (odd) permutation is an even (odd) permutation.

Solution : (i) Let p and q be even permutations in S_x where X is a set of n elements. Then, by definition of even permutations, p and q can be decomposed into products of even number of transpositions, say i and j so that

$$p = t_1 t_2 \ldots t_i \text{ and } q = u_1 u_2 \ldots u_j$$

where $t_1, \ldots t_i$ and $u_1 \ldots u_j$ are transpositions and i and j are even numbers. Thus their product

$$pq = (t_1 t_2 \ldots t_i)(u_1 u_2 \ldots u_j)$$

is expressible as a product of $(i + j)$ transpositions where $i + j$ is an even number. Hence pq is an even permutation. Thus *product of even permutations is a b.o.*

(ii) If in (i) above i and j are odd numbers then $i + j$ is an even number and the answer follows exactly on proceeding as in (i).

(iii) If in (i) above i is an even (odd) number and j is odd (even) then $i + j$ is an odd number. Now proceed as in (i).

(iv) Let p^{-1} denotes inverse of p, where p is an even permutation. Then $pp^{-1} = I$, where I is identity permutation, which is an even permutation. Thus pp^{-1} = even permutation, where p is even $\Rightarrow p^{-1}$ is an even permutation (by i). Similarly if $\overline{\alpha}^1$ is inverse of α, where α is an odd permutation, then

Permutation and Groups of Permutation

$\alpha \bar{\alpha}^1 = I$, where I is an even permutation being identity. Thus by (ii), α^{-1} is an odd permutation.

Problem 2.4 : Let A_x be set of all even permutations where X is any set of n elements. Then show that A_x forms a group of order $\frac{1}{2} \lfloor n$ under permutation multiplication.

Solution : We know that $A_x \subseteq S_x$ and $O(A_x) = \frac{1}{2} \lfloor n$, and that product of two even permutation is a b.o. in A_x. Associativity of elements of A_x follows from the fact that elements of A_x belong to S_x which is a group. The Identity permutation $I \in S_x$ as I is an even permutation and since, by Prob. 2.3 (iv), inverse of an even permutation is even thus each element of A_x has its inverse in A_x. Thus A_x is a group of order $\frac{1}{2} \lfloor n$.

Since, in general, permutation product does not obey commutative law therefore the group A_x is non-abelian.

Note : (1) If $X = \{1, 2, 3 \ldots n\}$, we know that A_x is denoted by A_n and S_x by S_n where $A_n \subseteq S_n$. Thus by Prob. 2.4, A_n is a *non-abelian group of order* $\frac{1}{2} \lfloor n$; and A_n is called the Alternating group. (It has nothing to do with the initials A.N. of the author!)

(2) The set $S_x - A_{x'}$ whose order is also $\frac{1}{2}\lfloor n$ and which is set of all odd permutations over set X, can not form a group as we have seen in Prob. 2.3 (ii) that the product of two odd permutations is even.

Problem 2.5 : Find a solution of the equation $ax = b$ in S_3 where $a = (1\ 2\ 3)$ and $b = (2\ 3)$ belong to S_3.

Solution : $ax = b$, where a and b are permutations from S_3, implies
$$x = a^{-1} b$$
$$= (1\ 2\ 3)^{-1} (2\ 3)$$

$$= (3\ 2\ 1)\ (2\ 3)$$

$$= \begin{pmatrix} 1 & 2 & 3 \\ 3 & 1 & 2 \end{pmatrix} \begin{pmatrix} 1 & 2 & 3 \\ 1 & 3 & 2 \end{pmatrix} = \begin{pmatrix} 1 & 2 & 3 \\ 2 & 1 & 3 \end{pmatrix}$$

This $x = (1\ 2)$, which is element μ_3 of S_3.

Problem 2.6 : Give an example to show that when cycles in the product into which a permutation is expressed are not disjoint then order of permutation is not equal to the LCM of lengths of the cycles.

Solution : Consider the elements $\mu_3 = \begin{pmatrix} 1 & 2 & 3 \\ 2 & 1 & 3 \end{pmatrix}$ and $\rho_1 = \begin{pmatrix} 1 & 2 & 3 \\ 2 & 3 & 1 \end{pmatrix}$ of the group S_3

Then $\mu_3 = (1\ 2)$ and $\rho_1 = (1\ 2\ 3)$
and
$$\mu_3 \rho_1 = (1\ 2)\ (1\ 2\ 3) = (1\ 3).$$

Now

$$O(\mu_3\ \rho_1) = O \begin{pmatrix} 1 & 2 & 3 \\ 3 & 2 & 1 \end{pmatrix} = 2 \text{ which is not equal to 6, the}$$
LCM of 2 and 3.

Review Exercise on Chapter 2

1. If $\alpha = (1\ 3\ 5)\ (1\ 2)$ and $\beta = (1\ 5\ 7\ 9)$ belong to S_9, find the product $\alpha^{-1}\beta\alpha$ and its order. Is $\alpha^{-1}\beta\alpha$ a cyclic permutation ?

2. Consider the permutations

$$p = \begin{pmatrix} 1 & 2 & 3 & 4 & 5 & 6 & 7 \\ 3 & 4 & 1 & 5 & 2 & 6 & 7 \end{pmatrix} \text{ from } S_7,$$

$$q = \begin{pmatrix} 1 & 2 & 3 & 4 & 5 & 6 & 7 & 8 \\ 8 & 2 & 6 & 3 & 7 & 4 & 5 & 1 \end{pmatrix} \text{ from } S_8,$$

and $\quad r = \begin{pmatrix} 1 & 2 & 3 & 4 & 5 & 6 & 7 & 8 & 9 & 10 \\ 4 & 2 & 5 & 8 & 6 & 1 & 3 & 7 & 9 & 10 \end{pmatrix}$

from S_{10}.
Find if $p \in A_7$, $q \in A_8$ and $r \in A_{10}$.

3. In S_7, let
$j = (1\ 2\ 3\ 5)$ and $k = (3\ 4\ 6\ 5)$.
Find (i) $(jk)^2$, (ii) $(jk)^{-2}$, and (iii) find if $(jk)^{-2} \in A_7$.

4. Let $a = (1\ 2\ 3\ 4\ 5) \in S_5$. Show that the set $\{a, a^2, a^3, a^4, a^5\}$ forms a group under permutation multiplication.

5. Distribute the elements of the set D_4 into even the odd permutations and find their orders.

6. Let $i = (1\ 2\ 3\ 4\ 5\ 6)$ and $j = (1\ 4)\ (2\ 5)\ (3\ 6)$ belong to S_6. Find order of the permutation ij and show that $O(ij) \neq$ LCM of lengths of the cycles i and j.

3 SUBGROUPS OF A GROUP

3.1 Subgroups

In the group table of permutation group S_3 (see 2.2.2 of chapter 2) we came across the concept of a group within a group. Out of total six elements of S_3, three elements namely ρ_0, ρ_1, ρ_2 form a group of their own under the same b.o. as that of the main group S_3; the identity element ρ_0 being common. Such 'smaller' groups within a group, if exist, are called subgroups of the group. Subgroups have a wonderful theory which contributes in understanding a group completely.

Definition 3.1.1 : A non-empty subset of a group (G, *) is called a *complex* of G.

For example the subsets $\{\rho_0\}, \{\rho_0, \rho_1\}, \{\rho_1, \rho_2\}$, *etc.* of S_3 = $\{\rho_0, \rho_1, \rho_2, \mu_1, \mu_2, \mu_3\}$ are complexes of the group S_3.

Definition 3.1.2 : A non-empty subset H (or a complex H) of a group (G, *) is called *subgroup of G* if the subset H itself forms a group of its own under the b.o. *.

We denote a subgroup H of a group (G, *) by writing H < G.

Thus H < (G, *) if

(*i*) H is a complex of G, and

(*ii*) (H, *) is a group.

The identity element *e* of the group (G, *) serves as identity element for all of the subgroups of G. Thus the id. element *e* of G is common between G and all of its subgroups.

Proper and Improper Subgroups : Two special subsets of G, namely the singleton {e} containing the id. element of G and the group G itself are always subgroups of G and are called *Improper* (or *Trivial*) subgroup of G. Other subgroups of G, if any, are called *Proper* (or *Non-Trivial*) subgroups. There are only two Improper subgroups of a group and all the groups have them. Notice that :

(i) if group G is finite, order of each of its proper subgroups will be less than the order of G, and

(ii) total number of subgroups of G will also be less than the order of group.

(iii) An infinite group may have an infinite number of subgroups.

Abelian Subgroups : A subgroup satisfying commutative law is an *abelian subgroup*. If a groups G is abelian, all of its subgroups will be abelian. Subgroups of a non-abelian group may also by abelian.

Note : (i) By a subgroup of a group we will always mean a proper subgroup unless mentioned otherwise.

(ii) The definition of a subgroup does not imply that all non-empty subsets of a group can form subgroups. For forming subgroups, the subsets should contain the identity elements e of group G and its elements should satisfy all the group axioms. For example, although $\mathbb{N}$ is a non-empty subset of the group $(\mathbb{R}, +)$, $\mathbb{N}$ is not a subgroup of $\mathbb{R}$ as the id. element of $\mathbb{R}$, wrt $* = +$, $0 \notin \mathbb{N}$. We then write $\mathbb{N} \not< \mathbb{R}$. Similarly $\mathbb{Q}^+ \not< (\mathbb{R}, +)$ even though $\mathbb{Q}^+ \subset \mathbb{R}$.

(iii) Since the b.o. $*$ of G is also the b.o. of its subgroup H, it implies that $*$ is a map from $H \times H$ into H or $* : H \times H \to H$ so that $*$ is a restriction of the map $* : G \times G \to G$ and thus $*$ is also called the *induced b.o. in H*.

3.1.1 Illustrative Examples

(A) *Subgroups of infinite groups* :

(i) Consider the group of integers $(\mathbb{Z}, +)$. For any non-zero integer n, *the set $n\mathbb{Z}$* forms a non-empty subset of $\mathbb{Z}$. Examples are :

$2\mathbb{Z} = \{ \ldots -6, -4, -2, 0, 2, 4, 6, \ldots \}$, which is set of all even integers,

$3\mathbb{Z} = \{\ldots -9, -6, -3, 0, 3, 6, 9, \ldots\}, \ldots,$

$(-4)\mathbb{Z} = \{\ldots 12, 8, 4, 0, -4, -8, -12, \ldots\} = 4\mathbb{Z}$, etc.

All of these non-empty subsets of $\mathbb{Z}$ form a group of their own under b.o. + so that they are subgroups of the group $(\mathbb{Z}, +)$, i.e., $2\mathbb{Z} < \mathbb{Z}, 3\mathbb{Z} < \mathbb{Z}, \ldots (-4)\mathbb{Z} < \mathbb{Z}, \ldots$.

In general, for any non-zero integer n, $n\mathbb{Z} < \mathbb{Z}$. When $n = 0$, $n\mathbb{Z}$, is trivial subgroup $\{0\}$ of $(\mathbb{Z}, +)$. Notice that $(\mathbb{Z}, +)$ and all its subgroups are abelian.

(ii) Since $\mathbb{Z} \subset \mathbb{Q}$ and $(\mathbb{Z}, +)$, $(\mathbb{Q}, +)$ both form groups, hence $\mathbb{Z} < \mathbb{Q}$. Similarly $(\mathbb{Z}, +) < (\mathbb{R}, +)$, $(\mathbb{Z}, +) < (\mathbb{C}, +)$. Also $(\mathbb{Q}, +) < (\mathbb{R}, +)$ and $(\mathbb{R}, +) < (\mathbb{C}, +)$ so that we have chain relationship in subgroups,

$(\mathbb{Z}, +) < (\mathbb{Q}, +) < (\mathbb{R}, +) < (\mathbb{C}, +)$.

In general, whenever $H < K < L < G$ then $H < L$ and $H < G$, a property of subgroups that can be proved easily.

(iii) The following infinite subsets of set $N_{2\times 2}(\mathbb{R})$ of all 2×2 invertible matrices form groups of their own under matrix multiplication :

$$H = \left\{ \begin{pmatrix} a & b \\ -b & a \end{pmatrix} : a, b \in \mathbb{R} \text{ and } a\,b \neq 0 \right\},$$

$$I = \left\{ \begin{pmatrix} a & b \\ b & a \end{pmatrix} : a, b \in \mathbb{R} \text{ and } a^2 - b^2 = 1 \right\},$$

and

$$J = \left\{ \begin{pmatrix} \cos\theta & -x\sin\theta \\ \frac{1}{x}\sin\theta & \cos\theta \end{pmatrix} : 0 \neq x, \theta \in \mathbb{R} \right\}$$

Thus H, I and J are subgroups of the group $v_{2\times 2}$ (ℝ) (or GL_2 (ℝ), i.e., H < GL_2 (ℝ), I < GL_2 (ℝ) and J < GL_2 (ℝ).

GL_2 (ℝ) is known to be a non-abelian group (as matrix multiplication, in general, does not commute). Check if the subgroups H, I and J are abelian.

(B) Subgroups of finite groups :

(i) Consider the group G = {–1, 1, i, –i}, where * = •, of fourth root of unity. Its non-empty subset including the id. elements 1 are

H_1 = {–1,1}, H_2 = {1, i}, H_3= {1, –i}, H_4 = {1, 1, i }, H_5 = {–1, 1, –i}, H_6 = {1, i, –i} and, of course, {1} and G itself. (The last two are improper subgroups of G). Out of first six complexes of G, only H_1 = {–1, 1} forms a groups of its own under * = •. For, in H_2 $i.i$ = –1 ∉ H_2 so that operation is not closed in H_2; in H_3 inv (–i) = i, as (–i) .i = 1, but i ∉ H_3; in H_4 (–1) i = –i ∉ H_4; in H_5 (–1) (–i) = i ∉ H_5 and in H_6, (i) (i) = –1 ∉ H_6. Thus the only proper subgroup of G is H_1 = {–1, 1}.

(ii) Consider two groups from the family of $\mathbb{Z}n$ -groups namely the groups ($\mathbb{Z}_3$, $+_3$) and ($\mathbb{Z}_4$, $+_4$), where

$\mathbb{Z}_3$ = {0, 1, 2}, $\mathbb{Z}_4$ = {0, 1, 2, 3}.

The complexes of $\mathbb{Z}_3$ including the id. element 0 are H_1 = {0, 1}, H_2 = {0, 2} and {0} and $\mathbb{Z}_3$ itself. You will easily find reasons why H_1 and H_2 are not forming groups of their own under * = $+_3$.

Thus the group $\mathbb{Z}_3$ has no (proper) subgroup at all. Similarly, by sorting out those complexes of $\mathbb{Z}_4$ which include 0 you will find that only H = {0, 2} is a subgroup of $\mathbb{Z}_4$. In chapter 5 you will come across Lagrange's powerful theorem which tells us that a finite group can have subgroup only when the order of the group is a divisible number. Thus like $\mathbb{Z}_3$, the groups $\mathbb{Z}_5$, $\mathbb{Z}_7$, $\mathbb{Z}_{11}$, ... do not have any proper subgroup (A finite group of order 2 can not have a proper subgroup also because it has no complexes, including its id. element, other than the two forming its improper subgroups.)

(iii) The complexes $H_1 = \{e, a\}$, $H_2 = \{e, b\}$ and $H_3 = \{e, c\}$ of Klein's 4-group $K_4 = \{e, a, b, c\}$ are its proper subgroups.

(iv) In the group table of the group S_3, the first block of the table containing the elements ρ_0, ρ_1 and ρ_2 forms a group table for the set $A_3 = \{\rho_0, \rho_1, \rho_2\}$. Hence $A_3 < S_3$. Other subgroups of S_3 are $\{\rho_0, \mu_1\}$, $\{\rho_0, \mu_2\}$ and $\{\rho_0, \mu_3\}$. Thus the group S_3 has four subgroups. From the properties of elements of S_3, we notice that all the four subgroups are abelian subgroups, while S_3 itself is a non-abelian group. Thus a non-abelian group may have abelian subgroups, showing that subgroups of a group may have additional properties not possessed by the group.

3.1.2 Criteria for subgroups

Obviously it is a tedious job to consider all possible complexes containing the id. element of a group for finding its proper subgroups. Instead, we apply one of the following criterias (necessary and sufficient conditions) to find proper subgroups of a given group.

Theorem 3.1.1 : Let $(G, *)$ be a group and H be a non-empty subset of G. Then H is a subgroup of G if and only if $a, b \in H \Rightarrow a * b^{-1} \in H$.

Proof : Given that $(G, *)$ is a group and H, a complex of G.

(i) *First part* : Assume that $H < G$, so that $b \in H \Rightarrow b^{-1} \in H$ and $a, b^{-1} \in H \Rightarrow a * b^{-1} \in H$, as the operation $*$ is closed in H.

(ii) *Second part* : Assume that for any $a, b \in H$, $a * b^{-1} \in H$ ...(1)

Since elements of H are drawn from group G, they satisfy associative law.

Since (1) is true for any a and b belonging to H, apply (1) on a, we get $a * a^{-1} \in H$, i.e., $e \in H$. Similarly applying (1) on e and b, we get $e * b^{-1} = b^{-1} \in H$, i.e., $b \in H \Rightarrow b^{-1} \in H$ so that identity e and inverses belong to H. Finally apply (1) in a and

Subgroups of a Group

b^{-1} we get $a, b^{-1} \in H \Rightarrow a * (b^{-1})^{-1} = a * b \in H$, so that $a * b \in H$, where $a, b \in H$ or * is a closed operation for H. Hence H < G.

Note : In the above theorem, if in particular * = + then H < G is and only if $a, b \in H \Rightarrow a - b \in H$, or if * = • then H < G if $a, b \in H \Rightarrow \frac{a}{b} \in H$.

Theorem 3.1.2 : Let H be a complex of a group (G , *). Then H < G if and only if

(i) $a \in H, b \in H \Rightarrow a * b \in H$

(ii) $a \in H \Rightarrow a^{-1} \in H$

Proof : Given the (G, *) is a group and H, a non-empty subset of G.

(i) *First part* : Assume that H < G so that (H, *) itself is a group and then (i) and (ii) hold in H.

(ii) *Second part* : Assume that the given conditions (i) and (ii) hold in H. To prove that H < G. Since H is a complex of G, elements of H satisfy associative law.

By condition (ii), $a^{-1} \in H$ so that by condition (i) $a, a^{-1} \in H \Rightarrow a * a^{-1} = e \in H$, *i.e.*, identity element belongs to H. Finally $a * a^{-1} = e, e \in H, \Rightarrow a^{-1}$ is inverse of $a \in H$ and by condition (ii) $a^{-1} \in H$ so that inverses belong to H. Hence elements of H satisfy all group conditions, *i.e.*, H < G.

Note : The above theorem 3.1.2 follows as a corollary to theorem 3.1.1.

Theorem 3.1.3 : A finite complex F of a group (G, *) is subgroup of G if and only if $a, b \in F \Rightarrow a * b \in F$.

Proof : Given that (G, *) is a group and F, a non-empty finite subset of G.

(i) *First part* : Assume that $F < G$ so that $(F, *)$ itself is a group and thus $a, b \in F \Rightarrow a * b \in F$ by closure property of $*$ in F.

(ii) *Second part* : Assume that for any $a, b \in F, a * b \in F$ ...(1)

By (1), $a * a = a^2 \in F$, $a^2 * a = a^3 \in F$, $a^3 * a = a^4 \in F$, ..., in general for any positive integer m, $a^m \in F$ so that $F = \{a, a^2, a^3, ..., a^m, ...\}$, which implies that F is an infinite set which is not true as F is given to be a finite subset of G. Thus some of the elements of F as given above are not distinct. Let for some $r, s \in \mathbb{Z}^+$ where $r > s$, $a^r = a^s \Rightarrow a^{r-s} = e$, where e is id. element of G. Since $r-s \in \mathbb{Z}^+$, $a^{r-s} \in F$ or $e \in F$ which shows the presence of e in F. Next, $r > s \Rightarrow r-s \geq 1 \Rightarrow r-s-1 \geq 0 \Rightarrow a^{r-s-1} \in F$. Now $a^{r-s-1} * a = e$, showing that a^{r-s-1} is inverse of the element $a \in F$. Finally elements in F obey associative law as $F \subset G$. Thus $F < G$.

Alternative proof of part (ii) : By (1), $a, b \in F \Rightarrow a * b \in F$ which implies that operation $*$ is a b.o. for F. Further since F is a finite subset of G, elements of F satisfy associative and cancellation laws so that F is a semi group in which elements satisfy cancellation law. Hence by Theorem 1.7.3, chapter 1, $F < G$.

3.1.3 Illustrative Examples

The above three criterias of subgroups are very useful in finding whether a complex of a group G is its subgroup. Here are a few examples :

(i) We know that $2\mathbb{Z} < (\mathbb{Z}, +)$. This can be checked by Theorem 3.1.1. Take $a = -2$ and $b = 6$, then $a * b^{-1} = (-2) + (6)^{-1} = (-2) + (-6) = -8$, which belongs to $2\mathbb{Z}$. Hence by Theorem 3.1.1., $2\mathbb{Z} < (\mathbb{Z}, +)$.

(ii) $3\mathbb{Z}$ is a subgroup of $(\mathbb{Z}, +)$ can be checked using Theorem 3.1.2. Take $a = -9$ and $b = 3$, then $a * b = (-9) + (3) = -6 \in 3\mathbb{Z}$ which is condition (i) of the theorem. Further, whenever $m \in 3\mathbb{Z}$, $m^{-1} \in 3\mathbb{Z}$ which is condition (ii) of the theorem. Hence by Theorem 3.1.2, $3\mathbb{Z} < (\mathbb{Z}, +)$.

(iii) The subsets $\{e, a\}$, $\{e, b\}$ and $\{e, c\}$ are non-empty finite subsets of Klein's groups K_4. By group table of K_4, $a * a = e$, $b * b = e$ and $c * c = e$, all the above subsets satisfy the condition of Theorem 3.1.3, hence they all are subgroups of K_4.

3.2 Lattice diagrams

A lattice diagram displays all the subgroups of a group. In the diagram subgroup is joined with the group by means of straight lines. The subgroup is always placed under the group, thus the main group (which is the largest) takes the top-most place and its improper subgroup $\{e\}$ (which is the smallest) takes the lower-most place in the diagram. Thus lattice diagrams look like family photographs of groups! See for yourself in the following examples :

Examples of lattice diagrams

(i) *The group* $(\mathbb{Z}_4, +_4)$: Its only proper subgroup is $\{0, 2\}$. $\{0\}$ and $\mathbb{Z}_4$ are the improper subgroups. Notice that $\{0\} < \{0, 2\}$.

$$\mathbb{Z}_4$$
$$|$$
$$\{0, 2\}$$
$$|$$
$$\{0\}$$

(ii) *The Klein's group* K_4 : It has three proper subgroups, namely, $\{e, a\}$, $\{e, b\}$ and $\{e, c\}$.

Notice also that $\{e\} < \{e, a\}$, $e < \{e, b\}$ and $\{e\} < \{e, c\}$.

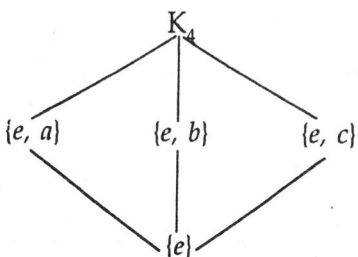

(iii) *The symmetric group* S_3. It has four proper subgroups, namely $A_3 = \{\rho_0, \rho_1, \rho_2\}$, $\{\sigma_0, \mu_1\}$, $\{\rho_0, \mu_2\}$ and $\{\rho_0, \mu_3\}$.

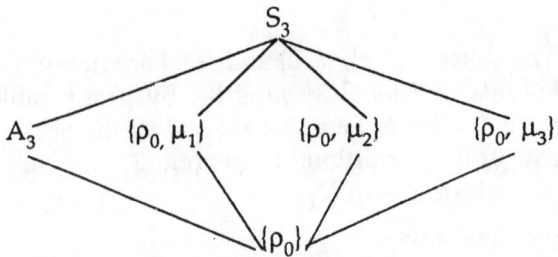

(vi) *The Octic group D_4*: D_4 has eight proper subgroups, namely $\{\rho_0, \nu_2, \tau_1, \tau_3\}$, $\{\rho_0, \nu_1, \nu_2, \nu_3\}$, $\{\rho_0, \nu_2, \tau_2, \tau_4\}$ (each of order 4), $\{\rho_0, \tau_1\}$, $\{\rho_0, \tau_2\}$, $\{\rho_0, \tau_3\}$ $\{\rho_0, \tau_4\}$ and $\{\rho_0, \nu_2\}$ (each of order 2). (You can apply Theorem 3.1.3 to check it.)

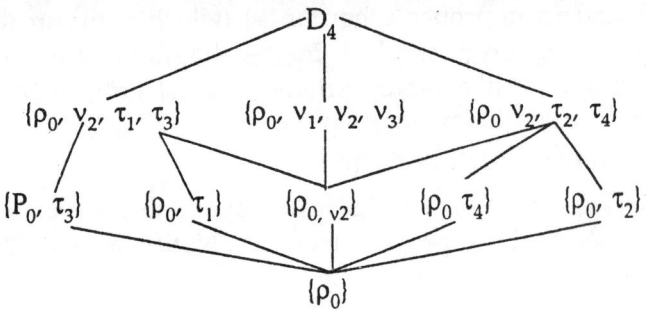

3.3 Algebra of subgroups

Given non-empty subsets H and K of a group $(G, *)$, we define the subsets H^{-1} and HK as follows :

Definition 3.3.1 : Let H be a non-empty subset of a group $(G, *)$. Then the set H^{-1} is the collection of all inverse elements of the element of H. Thus $H^{-1} = \{h^{-1} : h \in H\}$. Obviously $H^{-1} \neq \phi$. For example $\mathbb{N} = \{1, 2, 3, ...\}$ and $2\mathbb{Z} = \{... -4, -2, 0, 2, 4, ...\}$ are non-empty subsets of group $(\mathbb{Z}, +)$. Then

$\mathbb{N}^{-1}$ = set of all inverse elements of the elements of $\mathbb{N}$ (wrt $* = +$)

 $= \{-1, -2, -3, ...\}$, and similarly,

 $(2\mathbb{Z})^{-1} = \{... 4, 2, 0, -2, -4, ...\} = 2\mathbb{Z}$.

Definition 3.3.2 : Let H and K be non-empty subsets of a group $(G, *)$. Then the set HK is the collection of elements obtained by operating each element of H with every element of K. Thus HK is the 'product' set and is given by

Subgroups of a Group

HK = $\{h * k : h \in H$ and $k \in K\}$.

Similarly we define the product set KH as KH = $\{k * h : k \in K$ and $h \in H\}$

In particular, HH = $\{h * h : h \in H\}$. Also by definition 3.3.1, $(HK)^{-1} = \{(h * k)^{-1} : h \in H, k \in K\}$ and $(KH)^{-1} = \{(k * h)^{-1} : h \in H, k \in K\}$.

Examples

(1) Take two non-empty subsets

$$\mathbb{N} = \{1, 2, 3, \ldots\} \text{ and}$$
$$E = \{2, 4, 6, \ldots\} \text{ of group } (\mathbb{Z}, +).$$

Then $\mathbb{N}E = \{1 + 2, 1 + 4, 1 + 6, 2 + 2, 2 + 4, 2 + 6, \ldots\}$
$= \{3, 4, 5, 6, 7, 8, \ldots\}$,
$E\mathbb{N} = \{2 + 1, 2 + 2, 2 + 3, 4 + 1, 4 + 2, 4 + 3, \ldots\}$
$= \{3, 4, 5, 6, 7, \ldots\}$,

so that $\mathbb{N}E = E\mathbb{N}$ (Notice that $(\mathbb{Z}, +)$ is abelian)

Also, $\mathbb{N}^{-1} = \{-1, -2, -3, \ldots\}$.

$(\mathbb{N}E)^{-1}$ = set of inverse elements of the elements of the set $\mathbb{N}E$ (wrt $* = +$)
$= \{-3, -4, -5, -6, -7, -8, \ldots\}$

(2) Consider two non-empty subsets

$V_1 = \{a, b\}$ and $V_2 = \{e, a, b\}$ of Kleins group K_4. Then

$V_1^{-1} = \{a^{-1}, b^{-1}\} = \{a, b\} = V_1$

$V_2^{-1} = \{e^{-1}, a^{-1}, b^{-1}\} = \{e, a, b\} = V_2$

$V_1 V_2 = \{a * e, a * a, a * b, b * e, b * a, b * b\}$
$= \{a, e, c, b\}$,

$V_2 V_1 = \{e * a, e * b, a * a, a * b, b * a, b * b\}$
$= \{a, b, e, c\} = V_1 V_2$, and

$(V_1 V_2)^{-1} = \{a^{-1}, e^{-1}, c^{-1}, b^{-1}\} = \{a, e, c, b\} = V_1 V_2$.

Theorem 3.3.1 : Let H and K be non-empty subsets of a group (G, *). Then :

(a) $(HK)^{-1} = K^{-1} H^{-1}$
(b) $H < G \Leftrightarrow HH^{-1} = H$
(c) $H < G \Rightarrow HH = H$
(d) $H < G \Rightarrow H^{-1} = H$
(e) $HH = H \Rightarrow H < G$
(f) $H^{-1} = H \Rightarrow H < G$

Proof : Given that H and K are non-empty subsets of group (G, *).

(a) An arbitrary element of set $(HK)^{-1}$ is $(hk)^{-1}$, where $h \in H$ and $k \in K$

Since

$(hk)^{-1} = k^{-1} h^{-1}$, as H is subset of a group G and $k^{-1} \in K^{-1}$ and $h^{-1} \in H^{-1}$ so that $k^{-1} h^{-1} \in K^{-1} H^{-1}$ (by definition of product sets)

Thus $(hk)^{-1} \in (HK)^{-1} \Rightarrow (hk)^{-1} \in K^{-1} H^{-1}$
or $(HK)^{-1} \subseteq K^{-1} H^{-1}$ ...(i)

Conversely, $k^{-1} h^{-1} \in K^{-1} H^{-1}$
$\Rightarrow (hk)^{-1} \in K^{-1} H^{-1}$ as $k^{-1} h^{-1} = (hk)^{-1}$
But $(hk)^{-1} \in (HK)^{-1}$, thus $k^{-1} h^{-1} \in (HK)^{-1}$
so that $k^{-1} h^{-1} \in K^{-1} H^{-1} \Rightarrow k^{-1} h^{-1} \in (HK)^{-1}$, or
$K^{-1} H^{-1} \subseteq (HK)^{-1}$ ...(ii)

By (i) and (ii), $(HK)^{-1} = K^{-1} H^{-1}$, where H and K are non empty subsets of a group G.

(b) *First part :* Assume that $H < G$

An arbitrary element of set HH^{-1} is $h_1 h_2^{-1}$, where $h_1, h_2 \in H$.

Since $H < G$, $h_2 \in H \Rightarrow h_2^{-1} \in H$,
so that $h_1, h_2^{-1} \in H \Rightarrow h_1 h_2^{-1} \in H$ (by closure property of H)

Thus $h_1 h_2^{-1} \in HH^{-1} \Rightarrow h_1 h_2^{-1} \in H$, or $HH^{-1} \subseteq H$. Similarly $h \in H \Rightarrow h e^{-1} \in H$ ($H < G$)

Subgroups of a Group

But $he^{-1} \in HH^{-1}$ thus, $h \in HH^{-1}$ so that $h \in H \Rightarrow h \in HH^{-1}$, or $H \subseteq HH^{-1}$ Hence $HH^{-1} = H$, or $H < G \Rightarrow HH^{-1} = H$.

Second part : Assume that $HH^{-1} = H$ Then $HH^{-1} \subseteq H$, so that an element $h_1 \, h_2^{-1}$ of set HH^{-1}, where $h_1, h_2 \in H$, belongs to set H. Thus be Theorem 3.1.1, $H < G$, or

$$HH^{-1} = H \Rightarrow H < G.$$

(c) Given that $H < G$

An element of product set HH is $h_1 h_2$, where $h_1, h_2 \in H$.

Since $H < G$; $h_1, h_2 \in H \Rightarrow h_1 h_2 \in H$ *i.e.*, $HH \subseteq H$.

Conversely, $h \in H \Rightarrow h \, e \in H$ (as $H < G$)

But $h \, e \in HH$, thus $h \in HH$, so that $h \in H \Rightarrow h \in HH$, or $H \subseteq HH$

Hence $HH = H$, or $H < G \Rightarrow HH = H$

Converse of this result is not true (see part (*e*) of this theorem.

(d) Given that $H < G$

An element of set H^{-1} is h^{-1}, where $h \in H$.

Since $H < G$; $h \in H \Rightarrow h^{-1} \in H$

Thus $h^{-1} \in H^{-1} \Rightarrow h^{-1} \in H$, *i.e.*, $H^{-1} \subseteq H$.

Conversely, $h \in H \Rightarrow (h^{-1})^{-1} \in H$ (as $h \in H$, $H < G$ and in a group $h = (h^{-1})^{-1}$)

But $(h^{-1})^{-1} \in H^{-1}$ as $h^{-1} \in H$,

Thus $h \in H \Rightarrow h \in H^{-1}$, or $H \subseteq H^{-1}$

Hence $H < G \Rightarrow H^{-1} = H$.

Converse of this result is not true (see part (*f*) of this theorem).

(e) Consider subset

$$A = \{0, 2, 4, 6, ...\}$$

of the group $(\mathbb{Z}, +)$

Then $AA = \{0, 2, 4, 6, ...\} = A$

But $A \not< \mathbb{Z}$, as inverses for the elements of A are not available in A.

Thus $AA = A \not\Rightarrow A < (\mathbb{Z}, +)$,

which shows that, in general,

$HH = H \not\Rightarrow H < G$.

(f) Consider non-empty subset

$H = \{1, 3\}$ of the group $(\mathbb{Z}_4, +_4)$

From group table of $\mathbb{Z}_4$, inv. 1 = 3 and inv 3 = 1 so that $H^{-1} = \{1, 3\} = H$.

But $H \not< \mathbb{Z}_4$ as identity element $0 \notin H$. Thus $H^{-1} = H \not\Rightarrow H < G$, where G is any group.

Examples on Theorem 3.3.1

Let $G = \mathbb{Z}_4 = \{0, 1, 2, 3,\}$, $* = +_4$, and

$H = \{0, 2\}$, $T = \{0, 3\}$. Then $H^{-1} = \{0, 2\} = H$ and $T^{-1} = \{0, 1\} \neq T$. $HH^{-1} = \{0, 2\} = H$, $HH = \{0, 2\} = H$, $TT^{-1} = \{0, 1, 3\} \neq T$; $TT = \{0, 3, 2\}$. Thus $H < \mathbb{Z}_4$ but $T \not< \mathbb{Z}_4$.

Theorem 3.3.2 : Let H and K be any two subgroups of a group $(G, *)$. Then the product set HK is subgroup of G if and only if HK = KH.

Proof : Given that $(G, *)$ is a group and $H < G$, $K < G$.

(a) *First part* : Assume that $HK < G$. By Th. 3.1.1 (d),

$(HK)^{-1} = HK$

$\Rightarrow K^{-1} H^{-1} = HK$ \qquad (Th. 3.1.1. (a))

$\Rightarrow KH = HK$

Thus, $HK < G \Rightarrow HK = KH$

(b) *Second part* : Assume that HK = KH, where HK is a non-empty subset of group $(G, *)$.

In order to prove that $HK < G$, we will prove that $(HK)(HK)^{-1} = HK$, and in view of Theorem 3.3.1 (b), the result will follow.

Subgroups of a Group

$$\text{LHS} = (HK)(HK)^{-1}$$
$$= (HK)(K^{-1} H^{-1})$$
$$= H(KK^{-1})H^{-1} \quad \text{(By associative law)}$$
$$= H(K)H^{-1} \quad\quad (K < G)$$
$$= (HK)H^{-1}$$
$$= (KH)H^{-1}$$
$$= K(HH^{-1}) \quad\quad (H < G)$$
$$= KH = HK$$

Thus $HK = KH \Rightarrow HK < G$.

Corollary 3.3.1 : If H and K are subgroups of an abelian group (G, *) then HK < G.

Example on Theorem 3.3.2 : For Klein's group K_4, $H = \{e, a\}$ and $K = \{e, b\}$ are two subgroups. Then from group table of K_4, $HK = \{e, b, a, c\}$ and $KH = \{e, a, b, c\}$ so that $HK = KH$. Thus $HK < K_4$. (Here $HK = KH = K_4$, i.e., HK is improper subgroup of K_4.)

Theorem 3.3.3 : Intersection of any two subgroups of a group is its subgroup.

Proof : Let (G, *) be a group and H, K be two of its subgroups. In order to prove that $H \cap K < G$, we will prove that $a, b \in H \cap K \Rightarrow a * b^{-1} \in H \cap K$ and the result will follow in view of Theorem 3.1.1.

Now, $a, b \in H \cap K \Rightarrow a, b \in H$ and $a, b \in K$

$\Rightarrow a * b^{-1} \in H$ and $a * b^{-1} \in K$

(as $H < G, K < G$)

$\Rightarrow a * b^{-1} \in H \cap K$

Thus $a, b \in H \cap K \Rightarrow a * b^{-1} \in H \cap K$.

Note : (1) The above theorem can be extended for any number of subgroups of a group.

(2) Similar theorem is not true for the *union* of two or more subgroups of a group. For example, consider the subgroups $2\mathbb{Z}$, $3\mathbb{Z}$ and $4\mathbb{Z}$ of the group $(\mathbb{Z}, +)$.

Then
$$2\mathbb{Z} \cup 3\mathbb{Z} = \{..., -4, -3, -2, 0, 2, 3, 4, ...\}$$

and
$$2\mathbb{Z} \cup 4\mathbb{Z} = \{..., -6, -4, -2, 0, 2, 4, 6, ...\}$$

Then $2\mathbb{Z} \cup 3\mathbb{Z}$ is not subgroup of $(\mathbb{Z}, +)$ as the union is not closed wrt $* = +$, while $2\mathbb{Z} \cup 4\mathbb{Z}$ satisfies all group conditions and hence is a subgroup of $(\mathbb{Z}, +)$. (Notice that $2\mathbb{Z} \cup 4\mathbb{Z} = 2\mathbb{Z}$ and $4\mathbb{Z} \subset 2\mathbb{Z}$).

Thus union of two subgroups of a group may not be its subgroup.

(3) The following theorem provides us the condition under which union of subgroups becomes a subgroup.

Theorem 3.3.4 : The union of two subgroups of a group is subgroup if and only if one subgroup is contained in the other.

Proof: Let $(G, *)$ be a group and H, K be any two subgroups of G.

(a) *First part* : Assume that either $H \subseteq K$ or $K \subseteq H$ Then if $H \subseteq K$, $H \cup K = K$, which is a subgroup of G.

And if $K \subseteq H$, $H \cup K = H$, which is a subgroup G.

Thus if $H \subseteq K$ or $K \subseteq H$ then $H \cup K < G$.

(b) *Second part* : Assume that $H \cup K < G$. Let under this condition. $H \not\subseteq K$ or $K \not\subseteq H$.

Now $H \not\subseteq K \Rightarrow$ there exists h in H such that $h \notin K$ ...(i)

And $K \not\subseteq H \Rightarrow$ there exist k in K such that $k \notin H$ ...(ii)

Now $\qquad h \in H$ and $k \in K \Rightarrow h, k \in H \cup K$

$\Rightarrow \qquad hk \in H \cup K$ (as $H \cup K < G$)

⇒ $hk \in H$ or $hk \in K$

Let $hk \in H$

⇒ $h^{-1}(hk) \in H$ ($h \in H$, $H < G$ then $h^{-1} \in H$)

⇒ $(h^{-1}h)k \in H$

⇒ $k \in H$, contradicting (ii)

Then, let $hk \in K$

⇒ $(hk)k^{-1} \in K$ ($k \in K$, $K < G$ then $k^{-1} \in K$)

⇒ $h(kk^{-1}) \in K$

⇒ $h \in K$, contradicting (i)

Hence under the condition that $H \cup K < G$, $H \not\subseteq K$ or $K \not\subseteq H$ are not possible. Thus $H \cup K < G \Rightarrow$ either $H \subseteq K$ or $K \subseteq H$.

(The example given in Note 2 above illustrates this situation.)

3.4 Normalizer of an element and centre of a group

Definition 3.4.1 : Let $(G, *)$ be a group and $a \in G$. The set of all those elements of G which commute with the element a is called *normalizer* of a, denoted by $N(a)$.

Thus, $N(a) = \{x \in G : a * x = x * a\}$.

Obviously $a \in N(a)$.

Examples

(1) Since $a * e = e * a$, where e is id. element of G, $e \in N(a)$. Thus a and e always belong to $N(a)$ so that $N(a)$ is a non-empty subset of G.

(2) By definition, $N(e) = \{x \in G : e * x = x * e\}$.
$$= G$$

(3) If $G = (\mathbb{Z}, +)$ and $a = 2$ then since every integer commutes with 2 under $* = +$, i.e., $n + 2 = 2 + n$, for all $n \in \mathbb{Z}$, thus $N(2) = \mathbb{Z}$. In general if G is an abelian group and $a \in G$, $N(a) = G$.

(4) In the non-abelian groups S_3,

$N(\rho_1) = \{\rho_0, \rho_1, \rho_2\} = A_3$, as each of the elements ρ_0, ρ_1, ρ_2 commute with ρ_1.

Definition 3.4.2 : Let (G, *) be a group. The set of all those elements of G which commute with every element of G is called the centre of G denoted by Z(G) or simply by Z.

Thus, $Z(G) = \{c \in G : c * x = x * c$ for all $x \in G\}$ and $g \in Z(G)$ if $g \in G$ and g commutes with every element of G.

Note : (1) Since the id. element e of a group G commutes with every element of G, $e \in Z(G)$ and hence Z(G) is a non-empty subset of G.

(2) By definition, if $g \in Z(G)$ then $g \in G$ and $g x = x g$, for all $x \in G$. But $gx = xg$ for all $x \in G$ implies that $x \in N(g)$. So that $G \subseteq N(g)$. Since $N(g) \subseteq G$, thus $g \in Z(G) \Rightarrow N(g) = G$. Conversely $N(g) = G \Rightarrow G \subseteq N(g)$ so that $g \in G \Rightarrow g \in N(g)$ $\Rightarrow gx = xg$ for all $x \in G \Rightarrow g \in Z$. Thus we have the connection between $N(g)$ and $Z(G)$ as :

$g \in Z(G) \Leftrightarrow N(g) = G$.

(3) If G is a finite group, from above, $g \in Z(G) \Leftrightarrow$ $O\{N(g)\} = O\{G\}$

(4) Notice that N(a) is a property associated with an element a of a group G while Z(G) is property associated with the entire group. Hence N(a) is local and Z(G) is global in nature.

Examples : (1) For an abelian group G, Z(G) = G. Thus for example $Z(\mathbb{R}) = \mathbb{R}$ where $(\mathbb{R}, +)$ is known to be an abelian group.

(2) For non-abelian group S_3, $Z(S_3) - \{\rho_0\}$, as only ρ_0 commutes with every element of S_3.

Theorem 3.4.1 : If (G, *) is a group and $a \in G$, then $N(a) < G$.

Proof : By definition, $N(a) = \{x \in G : a * x = x * a\}$ and N(a) is non-empty subset of G.

Subgroups of a Group

Take two elements n_1, n_2 of $N(a)$. Then $n_1, n_2 \in G$ and $an_1 = n_1 a$, $an_2 = n_2 a$.

Now,

$a(n_1 n_2) = (an_1) n_2 = (n_1 a) n_2 = n_1 (an_2) = n_1(n_2 a) = (n_1 n_2) a$

or, $a(n_1 n_2) = (n_1 n_2) a \Rightarrow n_1 n_2 \in N(a)$, where $n_1, n_2 \in N(a)$.

Also,

$a n_1^{-1} = (n_1^{-1} n_1) a n_1^{-1} = n_1^{-1} (n_1 a) n_1^{-1} = n_1^{-1} (an_1) n_1^{-1} = (n_1^{-1} a) n_1 n_1^{-1} = n_1^{-1} a$

or, $a n_1^{-1} = n_1^{-1} a \Rightarrow n_1^{-1} \in N(a)$, where $n_1 \in N(a)$.

Thus we proved that $n_1, n_2 \in N(a) \Rightarrow n_1 n_2 \in N(a)$ and $n_1^{-1} \in N(a)$, i.e., $N(a) < G$.

Theorem 3.4.2 : If Z is centre of a group (G, *) then $Z < G$.

Proof : By definition,

$Z = \{c \in G : c * x = x * c$ for all $x \in G\}$ and Z is a non empty subset of G. Take two elements g_1, g_2 of Z. Then $g_1, g_2 \in G$ and for all $x \in G$,

$$g_1 x = xg_1 \text{ and } g_2 x = xg_2.$$

Now,

$$g_2^{-1} x = g_2^{-1} x (g_2 g_2^{-1}) = g_2^{-1} (xg_2) g_2^{-1}$$

$$= g_2^{-1} (g_2 x) g_2^{-1}$$

$$= (g_2^{-1} g_2) x g_2^{-1}$$

$$= x g_2^{-1}$$

i.e., $g_2^{-1} x = x g_2^{-1}$ for all $x \in G$

$\Rightarrow \quad g_2^{-1} \in Z$

Consider

$(g_1 g_2^{-1}) x$, where $g_1, g_2^{-1} \in Z$ and x is any element of G

$$= g_1 (g_2^{-1} x)$$
$$= g_1 (x \ g_2^{-1}) = (g_1 x) g_2^{-1} = (x g_1) g_2^{-1} = x(g_1 g_2^{-1})$$

or, $(g_1 g_2^{-1})x = x(g_1 \ g_2^{-1})$, for any $x \in G$,,

$\Rightarrow \quad g_1 g_2^{-1} \in Z$

Thus $Z < G$.

3.5 Group Actions and Stabilizers

In a group (G, 0) we come across the 'products' of elements of the types $g_1 \cdot g_2$ or $(g_1 \cdot g_2) \cdot g_3$ etc. or by dropping the operation 0, the products $g_1 g_2$, $(g_1 g_2) g_3$ where all the elements g_1, g_2, g_3 are 'internal', i.e., all of them belong to G and their products also belong to G. We now define the 'products' of elements of G with the elements of a non empty set A, not necessarily contained in G, such that the new element belongs to A. For each pair (a, g) of elements obtained by combining a and g is written as $a * g$ such that $a * g \in A$ and we interpret the element $a * g$ by saying that the element g has 'acted' on element a. The formal definition of G acting on A is as follows :

Definition 3.5.1 : Let (G, 0) be a group in which the product $g_1 \cdot g_2$ of elements g_1 and $g_2 \in G$ is denoted by $g_1 g_2$ and let A be a non-empty set not necessarily a subset of G. We say that *group G acts on set A by the product* * (or that there is *action_* of G on A) if for each pair of element (a, g) where $a \in A$ and $g \in G$, there exists an element $a * g \in A$ satisfying following two conditions :

(I_A) $a * e = a$, for all $a \in A$

(II_A) $(a * g_1) * g_2 = a * (g_1 g_2)$; $g_1, g_2, g_1 g_2 \in G$.

Note : (1) Thus when a group acts on a set A, there exists an operation * such that element g of G, when acted on elements of A, changes element a of A to another element $a * g$ of A, where $a * g$ satisfies conditions I_A and II_A. So in a group action, elements of G change elements of a non-empty set A to another elements of A.

(2) The condition I_A implies that the identity element e when acts on elements of A leave them unchanged. II_A implies

Subgroups of a Group

that the effect of g_1 acting first on a and then g_2 acting is same as that of element $g_1 g_2$ acting on a.

(3) The concept of action $*$ of G on A is similar to the external binary operation $* : A \times G \to A$. In the general theory group actions play a significant role and provide us a way of obtaining subgroups of groups.

3.5.1 Illustrative Examples

(1) The action $*$ of group $(\mathbb{R}, +)$ on the set $\mathbb{R}^2$ of all ordered pairs of elements of $\mathbb{R}$ is by the operation $*$ defined by $(a, b) * g = (a + gb, b)$, for all $(a, b) \in \mathbb{R}^2$ and $g \in (\mathbb{R}, +)$.

For the group $(\mathbb{R}, +)$, $e = O$ so that $(a, b) * 0 = (a + ob, b) = (a, b)$ which is I_A.

For any $g_1, g_2 \in (\mathbb{R}, +)$ and $(a, b) \in \mathbb{R}^2$, $(a, b) * g_1) * g_2$
$$= (a + g_1 b, b) * g_2$$
$$= (a + g_1 b + g_2 b, b) = (a (g_1 + g_2) b, b)$$
$$= (a, b) * (g_1 + g_2)$$
$$= (a, b) * (g_1 g_2),$$

which is II_A.

(2) A group G acts on itself (i.e., A = G) by the operation $*$, defined as
$$a * g = a^g = g^{-1} ag, \; a, g \in G.$$
Then
$$a * e = a^e = e^{-1} ae = a \text{ for all } a \in G, \text{ which is } I_A.$$
$$(a * g_1) * g_2 = \left(a^{g_1}\right)^{g_2} = (g_1^{-1} a g_1) * g_2$$
$$= g_2^{-1} (g_1^{-1} a g_1) g_2$$
$$= g_2^{-1} g_1^{-1} a g_1 g_2 = (g_1 g_2)^{-1} a (g_1 g_2) = a * g_1 g_2, \text{ which is } II_A.$$

(3) A subgroup H of group G acts on G when action $*$ is defined as $a * g = ag$, where the product ag is as defined in G.

Then

$a * e = ae = a$ for all $a \in H$, which is I_A. Also, for any g_1, $g_2 \in G$ and $a \in H$,

$$(a * g_1) * g_2 = (ag_1) * g_2 = (ag_1) g_2 = a (g_1 g_2)$$
$$= a * (g_1 g_2),$$

which is II_A.

Groups acting on a set provide us the definitions of *orbit* and *stabilizer* of an element.

Definition 3.5.2 : Let a group G acts on a non-empty set A and let $a \in A$. The *orbit* of element a, denoted by orb (a), is the set of all those elements of A which are obtained by action of some elements of G on a.

Thus

orb $(a) = \{b \in A : b = a * g$, for some $g \in G\}$ and orb $(a) \subset A'$.

Definition 3.5.3 : Let a group G acts on a non-empty set A and $a \in A$. Then *stabilizer* of element a, denoted by stab (a), is the set of all those elements of G which when acted on a leave a unchanged.

Thus

Stab $(a) = \{g \in G : a * g = a\}$ and stab $(a) \subset G$.

Examples

(1) When a group G acts on itself the action * is given by $a * g = g^{-1} a g$, $a, g \in G$. Thus for any $b \in G$,

Stab $(b) = \{g \in G : b * g = b\}$

The condition $b * g = b$

$\Rightarrow \quad g^{-1} b g = b$

$\Rightarrow \quad g (g^{-1} bg) = gb$

$\Rightarrow \quad bg = gb$

Thus

Stab $(b) = \{g \in G : bg = gb\}$

$= Z(G)$, the center of group G.

(2) When a subgroup H of a group G acts on group G, the action $*$ is given by $a * g = ag$, where $a \in H, g \in G$.

Then stab $(g) = \{e\}$ as only $e \in H$ such that $g * e = g$.

(3) When the symmetric group S_3 acts on the set $A = \{1, 2, 3\}$ by permuting the vertices of an equilateral triangle, 1 is left unchanged in two of the elements of S_3, namely

$$\rho_0 = \begin{pmatrix} 1 & 2 & 3 \\ 1 & 2 & 3 \end{pmatrix} \text{ and } \mu_1 = \begin{pmatrix} 1 & 2 & 3 \\ 1 & 3 & 2 \end{pmatrix} \text{ so that stab } (1) = \{\rho_0, \mu_1\}.$$

Similarly stab $(2) = \{\rho_0, \mu_2\}$ and stab $(3) = \{\rho_0, \mu_3\}$.

Theorem 3.5.1 : Let a group G acts on a non-empty set A by $*$ and $a, b \in A$ such that $a * g = b$ where $g \in G$ then $a = b * g^{-1}$.

Proof : Since G is a group, $g \in G \Rightarrow g^{-1} \in G$

Thus $\quad a * g = b$

$\Rightarrow (a * g) * g^{-1} = b * g^{-1}$

$\Rightarrow \quad a * (gg^{-1}) = b * g^{-1}$ (Applying II_A in LHS)

$\Rightarrow \quad\quad a * e = b * g^{-1}$ (by I_A)

$\Rightarrow \quad\quad a = b * g^{-1}$.

Theorem 3.5.2 : Let a group G acts on a non-empty set A by the action $*$ and $a \in A$ then stab $(a) < G$.

Proof : Let $b, c \in$ stab (a). Then by definition $b, c \in G$ and

$$a * b = a, \quad a * c = a$$

so that

$$a * bc = (a * b) * c = (a) * c = a,$$

or $\quad\quad a * bc = a$

$\Rightarrow \quad\quad bc \in$ stab (a)

Thus $\quad b, c \in$ stab $(a) \Rightarrow bc \in$ stab (a) ...(1)

Next,

$$b \in \text{stab } (a) \Rightarrow a * b = a$$

$$\Rightarrow a = a * b^{-1} \qquad \text{(by The. 3.5.1)}$$
$$\Rightarrow b^{-1} \in \text{stab}(a)$$

Thus $b \in \text{stab}(a) \Rightarrow b^{-1} \in \text{stab}(a)$ ...(2)

Hence by (1) and (2), stab $(a) < G$.

Solved Problems Set

Problem 3.1 : If H is subgroup of a group (G, *), then prove that :

(i) the identity of H is same as that of G; and

(ii) the inverse of any element h of H is same as the inverse of h regarded as an element of G.

Solution : Given (G, *) is a group and $H < G$

(i) Let identity of G be e and identity of H be e'. Then for any $a \in H$, $a * e' = a$ —(1) (as H is a group) and e' is identity of H.

$$a \in H \Rightarrow a \in G$$
$$\Rightarrow a \in e = a - (2) \qquad \text{(as } e \text{ is identity of G)}$$

By (1) and (2),
$$a * e' = a * e \Rightarrow e' = e \text{ (by LCL)}$$

(ii) Let e be the id. element of G. Then by (i), e is id-element of H. For any $h \in H$, let h_1 be inverse of h and when $h \in G$, let h_2 be inverse of h

Then
$$h * h_1 = e, \text{ and}$$
$$h * h_2 = e$$

so that
$$h * h_1 = h * h_2$$
$$\Rightarrow h_1 = h_2 \qquad \text{(by LCL).}$$

Problem 3.2 : If $H < G$ and $x \in G$, where (G, *) is a group then show that the set $H^x = \{x^{-1} * h * x : h \in H\}$ is a subgroup of G.

Solution : Writing $x^{-1} * a * x$ as $x^{-1}ax$ and $x^{-1} * b * x$ as $x^{-1}bx$ for any $a, b, x \in G$ and denoting them respectively by a^x and b^x, we have

Subgroups of a Group

$$a^x b^x = (x^{-1} ax)(x^{-1} bx)$$
$$= x^{-1} a (xx^{-1}) bx = x^{-1} (ab) x = (ab)^x$$

i.e,
$$a^x b^x = (ab)^x \qquad \ldots(1)$$

and
$$(a^x)^{-1} = (x^{-1} ax)^{-1} = x^{-1} (x^{-1} a)^{-1}$$
$$= x^{-1} (a^{-1} (x^{-1})^{-1}) = x^{-1} a^{-1} x$$
$$= (a^{-1})^x$$

i.e.,
$$(a^x)^{-1} = (a^{-1})^x \qquad \ldots(2)$$

In order to show that $H^x < G$, take two arbitrary elements h_1^x and h_2^x of H^x so that $h_1^x = x^{-1} h_1 x$ and $h_2^x = x^{-1} h_2 x$, where

$$h_1, h_2 \in H$$

Then $\qquad h_1^x h_2^x = (h_1 h_2)^x \qquad$ (by 1)

and $\qquad (h_1^x)^{-1} = \left(h_1^{-1}\right)^x \qquad$ (by 2)

But $\qquad h_1 h_2 \in H \Rightarrow h_1^x h_2^x \in H^x$, and

$$h_1^{-1} \in H \Rightarrow (h_1^x)^{-1} \in H^x. \text{ Thus } H^x < G.$$

(*Note* : The subgroup H^x is known as a *conjugate subgroup* to H in G. We will return to H^x in chapter 6.)

Problem 3.3 : If S_n is a group of all permutations over first n natural numbers and A_n is set of all even permutations, prove that

(*i*) $A_n < S_n$

(*ii*) $A_3 < S_3$.

Solution : (*i*) Since S_n is a finite group of order $\lfloor n$ and A_n is a finite subset of S_n whose order is $\frac{1}{2} \lfloor n$. Let $\alpha, \beta \in A_n$ so that α and β are even permutations over 1, 2, ... n. From chapter 2, we know that the product of two even permutations is also an even permutation, hence $\alpha\beta \in A_n$.

Thus $\alpha, \beta \in A_n \Rightarrow \alpha\beta \in A_n$, by Th. 3.1.3, $A_n < S_n$.

(ii) $A_3 < S_3$ follows form (i) above, on putting $n = 3$. Alternatively, since $\rho_1, \rho_2 \in A_3$ where

$$\rho_1 = \begin{pmatrix} 1 & 2 & 3 \\ 2 & 3 & 1 \end{pmatrix} \text{ and } \rho_2 = \begin{pmatrix} 1 & 2 & 3 \\ 3 & 1 & 2 \end{pmatrix}$$

then,
$$\rho_1 \rho_2^{-1} = \begin{pmatrix} 1 & 2 & 3 \\ 2 & 3 & 1 \end{pmatrix} \begin{pmatrix} 3 & 1 & 2 \\ 1 & 2 & 3 \end{pmatrix}$$

$$= \begin{pmatrix} 1 & 2 & 3 \\ 3 & 1 & 2 \end{pmatrix} = \rho_2 \in A_3$$

Thus

$\rho_1, \rho_2, \in A_3 \Rightarrow \rho_1 \rho_2^{-1} \in A_3$, by Th. 3.1.1, $A_3 < S_3$.

Problem 3.4 : Let (G, ∗) be a group, H < G and for $x \in$ G, x H and Hx denote, respectively, the sets

x H = $\{x * h : h \in H\}$ and

Hx = $\{h * x : h \in H\}$

then prove that the set

N = $\{x \in G : xH = Hx\}$ forms a subgroup of G.

Solution : Let $n_1, n_2 \in N$. Then $n_1, n_2 \in G$ and $n_1 H = Hn_1$ and $n_2 H = H n_2$.

Now $\qquad n_2 \in N$

$\Rightarrow \qquad n_2 H = Hn_2 \Rightarrow n_2^{-1} (n_2 H) n_2^{-1} = n_2^{-1} (Hn_2) n_2^{-1}$

$\Rightarrow (n_2^{-1} n_2) H\ n_2^{-1} = (n_2^{-1} H (n_2\ n_2^{-1})$

$\Rightarrow \qquad Hn_2^{-1} = n_2^{-1} H$

$\Rightarrow \qquad n_2^{-1} \in N$

i.e.,

$n_2 \in N \Rightarrow n_2^{-1} \in N$

Further
$$n_1\, n_2^{-1}\, H = n_1\, (n_2^{-1}\, H)$$
$$= n_1(H\, n_2^{-1}) \quad \text{(as } n_2^{-1} \in N\text{)}$$
$$= (n_1\, H)\, n_2^{-1} \quad (H < G)$$
$$= (H\, n_1)\, n_2^{-1} \quad \text{(as } n_1 \in N\text{)}$$
$$= H\, (n_1\, n_2^{-1})$$

i.e.,
$$(n_1\, n_2^{-1})\, H = H\, (n_1\, n_2^{-1}) \Rightarrow n_1\, n_2^{-1} \in N$$

or, $\quad n_1, n_2 \in N \Rightarrow n_1\, n_2^{-1} \in N.$ Thus $N < G.$

Problem 3.5 : Let $(G, *)$ be a group and $g \in G.$ Prove that
(i) the set $H = \{x \in G : x * g^2 = g^2 * x\}$ is subgroup of G
and
(ii) the set $K = \{x \in G : x * g = g * x\}$ is subgroup of $H.$

Solution : (i) For any $x, g \in G$, denote $x * g^2$ by $x\, g^2$, $g^2 * x$ by $g^2\, x$

By definition $H \neq \phi$ as atleast $e \in H$, because $e \in G$ and $e g^2 = g^2 e = g^2$

Let $\qquad h_1, h_2 \in H$ so that
$$h_1\, g^2 = g^2\, h_1 \text{ and } h_2\, g^2 = g^2\, h_2$$
Now $\qquad h_2 \in H \Rightarrow h_2 \in G$, and $h_2\, g^2 = g^2\, h_2$
$\Rightarrow \qquad h_2^{-1}\, (h_2\, g^2)\, h_2^{-1} = h_2^{-1}\, (g^2\, h_2)\, h_2^{-1}$
$\Rightarrow \qquad\qquad g^2\, h_2^{-1} = h_2^{-1}\, g^2$
$\Rightarrow \qquad\qquad h_2^{-1} \in H$

Thus $\qquad h_2 \in H \Rightarrow h_2^{-1} \in H.$
Further
$$h_1\, h_2^{-1} g^2 = h_1\, (h_2^{-1}\, g^2)$$

$$= h_1 (g^2 \; h_2^{-1}) \text{ as } h_2^{-1} \in H$$
$$= (h_1 \; g^2) \; h_2^{-1} = g^2 \; h_1 \; h_2^{-1}$$

i.e.,
$$h_1 \; h_2^{-1} \; g^2 = g^2 \; h_1 \; h_2^{-1} \Rightarrow h_1 \; h_2^{-1} \in H$$

or
$$h_1, h_2 \in H \Rightarrow h_1 \; h_2^{-1} \in H. \text{ Thus } H < G.$$

(ii) Let $j \in K$ then by definition of K, $j \in G$ and
$$jg = gj$$
$$\Rightarrow \quad (jg) \; g = (gj) \; g$$
$$\Rightarrow \quad jg^2 = g \; (jg)$$
$$= g \; (gj)$$
$$= g^2 \; j$$

i.e.,
$$jg^2 = g^2 \; j \Rightarrow j \in H$$
Thus $\quad j \in K \Rightarrow \quad j \in H$

$\Rightarrow K \subsetneq H$. Also K is non-empty as atleast $e \in K$ because $e \in G$ and $eg = ge$.

Now proceed exactly as in (i) above to prove that $K < H$.

Problem 3.6 : Let H be the set of all 2×2 matrices A with their elements from $\mathbb{R}$ such that $A' = A^{-1}$. Then prove that $H < GL_2 \; (\mathbb{R})$ where $GL_2 \; (\mathbb{R})$ is general linear group.

Solution : $GL_2 \; (\mathbb{R})$ is group of all 2×2 invertible matrices with their elements from R under $* = $ matrix multiplication.

Then given H is a subset of $GL_2 \; (\mathbb{R})$. Since $I_2 = \begin{pmatrix} 1 & 0 \\ 0 & 1 \end{pmatrix} \in H$, H is non-empty.

Let $A, B \in H$ so that $A' = A^{-1}$ and $B' = B^{-1}$.

Now by property of transpose of a matrix we know that
$$(AB)^1 = (AB') \; B^1 A^1 = B^{-1} \; A^{-1} = (AB)^{-1}$$

Subgroups of a Group 145

$\Rightarrow$ AB $\in$ H. Thus A, B $\in$ H $\Rightarrow$ (AB) $\in$ H

Finally, if A $\in$ H then since

$$(A^{-1})' = (A')' \quad \text{as } A^{-1} = A'$$
$$= A$$
$$= (A^{-1})^{-1}$$

$\Rightarrow \qquad A^{-1} \in H$, *i.e.*, A $\in$ H $\Rightarrow A^{-1} \in$ H

Thus $\qquad H < GL_2 (\mathbb{R})$.

(*Note* : The matrices satisfying $A' = A^{-1}$ are called orthogonal matrices and the subgroup H is called 2 by 2 orthogonal group.

Problem 3.7 : Let H be the set of all $n \times n$ matrices A with their elements from $\mathbb{R}$ such the determinant, $|A| = 1$. Then prove that for $n \geq 1$, $H < G L_n (\mathbb{R})$.

Solution : Since $|A| \neq O$, A is inevrtible *i.e.*, $A \in GL_n (\mathbb{R})$. Thus

$H = \{ A \in GL_n (\mathbb{R}) : |A| = 1\}$ is a subset of $GL_n (\mathbb{R})$. Also H is non-empty as

$$I_n = \begin{pmatrix} 1 & 0 & \dots & 0 \\ 0 & 1 & \dots & 0 \\ \dots & \dots & \dots \\ 0 & 0 & \dots & 1 \end{pmatrix}_{n \times n} \in H$$

Let B, C $\in$ H then B, C $\in GL_n (\mathbb{R})$ and $|B| = 1$, $|C| = 1$. By a property of determinants,

$|BC| = |B| |C| = 1 \Rightarrow BC \in H$

Thus B, C $\in$ H $\Rightarrow$ BC $\in$ H ...(1)

Further, for any B $\in$ H,

$|B| |B^{-1}| = |BB^{-1}| = |I_n| = 1$

$\Rightarrow |B^{-1}| = 1/|B| = 1 \Rightarrow B^{-1} \in H$

i.e.,

B $\in$ H $\Rightarrow B^{-1} \in$ H ...(2)

By (1) and (2), $H < GL_n (\mathbb{R})$.

(*Note* : This subgroup H of general linear group GLn (ℝ) is called *Special linear group* and is denoted $SL_n(\mathbb{R})$.)

Problem 3.8 : Show that the subset

$$H = \left\{ \begin{pmatrix} 1 & r \\ 0 & 1 \end{pmatrix} : r \in \mathbb{R} \right\}$$

of general linear group $GL_2(\mathbb{R})$ is an abelian subgroup of $GL_2(\mathbb{R})$.

Solution : Let A, B ∈ H.

Then by definition of H,

$$A = \begin{pmatrix} 1 & x \\ 0 & 1 \end{pmatrix}, \text{ and } B = \begin{pmatrix} 1 & y \\ 0 & 1 \end{pmatrix}$$

where $x, y \in \mathbb{R}$. Since

$$AB = \begin{pmatrix} 1 & x+y \\ 0 & 1 \end{pmatrix} \text{ and } x + y \in \mathbb{R},$$

AB ∈ H.

Thus A, B ∈ H ⇒ AB ∈ H ...(1)

Further for $A = \begin{pmatrix} 1 & x \\ 0 & 1 \end{pmatrix} \in H$, where $x \in \mathbb{R}$,

$$A^{-1} = \begin{pmatrix} 1 & x \\ 0 & 1 \end{pmatrix}^{-1} = \begin{pmatrix} 1 & -x \\ 0 & 1 \end{pmatrix}. \text{ Since } -x \in \mathbb{R},$$

$A^{-1} \in H$, i.e.,

A ∈ H ⇒ A^{-1} ∈ H ...(2)

Thus by (1) and (2), $H < GL_2(\mathbb{R})$.

Also, $AB = \begin{pmatrix} 1 & x+y \\ 0 & 1 \end{pmatrix} = \begin{pmatrix} 1 & y+x \\ 0 & 1 \end{pmatrix} = BA$, for any A, B ∈

H, therefore H is abelian.

Problem 3.9 : Let H and K be finite subgroups of a group G, then prove that

$$O(HK) = \frac{O(H)\, O(K)}{O(H \cap K)}$$

Subgroups of a Group

In particular if $H \cap K = \{e\}$ then
$O(HK) = O(H) O(K)$

Solution : Let $H = \{h_1, h_2, \ldots h_m\}$ and $K = \{k_1, k_2, \ldots, k_n\}$ so that $O(H) = m$ and $O(K) = n$. Then by definition
$$HK = \{h_i k_j : 1 \leq i \leq m, 1 \leq j \leq n\}$$
and

$O(HK)$ = Number of distinct elements $h_i k_j$ in the set HK.

In above set HK, all the elements $h_i k_j$ may not be distinct. We will first show that an element, say $h_1 k_1$ of set HK where $h_1 \in H$ and $k_1 \in K$ is repeated $O(H \cap K)$ times. For some i and j, let

$$h_1 k_1 = h_i k_j$$
$\Rightarrow \quad h_i^{-1} (h_1 k_1) k_1^{-1} = h_i^{-1} (h_i k_j) k_1^{-1}$

$\Rightarrow \quad h_i^{-1} h_1 = k_j k_1^{-1} = l$, say where $l \in H \cap K$

$\Rightarrow \quad h_i = h_1 e^{-1}$ and $k_j = l k_1$

so that, $\quad h_1 k_1 = h_i k_j = (h_1 l^{-1}) (l k_1), l \in H \cap K$

Similarly for some $t \in H \cap K$,
$$h_1 k_1 = (h_1 t^{-1})(t k_1), \ldots, \text{etc}$$

Thus the element $h_1 k_1$ is repeated $O(H \cap K)$ times in the set HK. Likewise all the elements on set HK repeat themselves $O(H \cap K)$ times

Thus, Number of distinct elements in set

$$HK = \frac{O(H) O(K)}{O(H \cap K)}$$

i.e., $\quad O(HK) = \dfrac{O(H) O(K)}{O(H \cap K)}.$

In particular, if $H \cap K = \{e\}$, $O(H \cap K) = 1$, so that $O(HK) = O(H) O(K)$.

(For alternative proof, see problem 5.7 in chapter 5)

Problem 3.10: Let H and K be subgroups of a finite group G, where $O(G) = n$, such that $O(H) > \sqrt{n}$ and $O(K) > \sqrt{n}$ then prove that $O(H \cap K) > 1$.

Solution: Since $HK \subseteq G$,

$$O(G) \geq O(HK) = \frac{O(H) O(K)}{O(H \cap K)} \text{ (by Prob. 3.9)}$$

i.e.,

$$O(G) \geq \frac{O(H) O(K)}{O(H \cap K)} \qquad ...(1)$$

Now since $O(H) > \sqrt{n}$ and $O(K) > \sqrt{n}$, therefore

$$\frac{O(H) O(K)}{O(H \cap K)} > \frac{\sqrt{n} \sqrt{n}}{O(H \cap K)} = \frac{n}{O(H \cap K)} = \frac{O(G)}{O(H \cap K)}$$

So, by (1)

$$O(G) \geq \frac{O(H) O(K)}{O(H \cap K)} > \frac{O(G)}{O(H \cap K)},$$

or $\quad O(G) > \dfrac{O(G)}{O(H \cap K)} \Rightarrow O(H \cap K) > 1.$

Problem 3.11: Apply subgroup criteria to show that the subsets $V = \{\rho_0, v_2, \tau_1, \tau_3\}$, $H_1 = \{\rho_0, l, m_1\}$, $H_2 = \{\rho_0, l, m_4\}$, $H_3 = \{\rho_0, l_2, m_2\}$ and $H_4 = \{\rho_0, l_3, m_3\}$ are subgroups of alternating group A_4 and that V is Klein's group.

Solution: Take any two elements v_2 and τ_3 of V. By group table of A_4 (or, by direct calculation), Inverse of $\tau_3 = \tau_3$ and $v_2 \tau_3^{-1} = v_2 \tau_3 = \tau_1 \in V$. Thus $V < A_4$. Also, inverse of $m_1 = l$ and $l\, m_1^{-1} = ll = m_1 \in H_1$ thus $H_1 < A_4$. Similarly you can show that $H_2 < A_4$ and $H_3 < A_4$.

Since $v_2^2 = v_2$, $\tau_1^2 = \tau_1$, $\tau_3^2 = \tau_3$ and also $v_2 \tau_1 = \tau_1 v_2 = \tau_3$, $v_2 \tau_3 = \tau_3 v_2 = \tau_1$, $\tau_1 \tau_3 = \tau_3 \tau_1 = v_2$, therefore the subgroup $V = \{\rho_0, v_2, \tau_1, \tau_3\}$ of the group A_4 is Klein's group K_4.

Subgroups of a Group

Problem 3.12 : Let H be the set of all rotations in the plane about the origin, given by

$H = \{rot_\theta : \theta \in \mathbb{R}\}$ and O_2 be the orthogonal group, then show that $H < O_2$.

Solution : For orthogonal group O_2, see 2.3.1 in chapter 2. We will apply Theorem 3.1.2 to prove that given H is a subgroup of O_2.

Atleast $rot_0 \in H$ so that H is a complex of O_2. 'Product' of two rotations in the plane about origin is closed, as if $\theta, \phi \in \mathbb{R}$ then $rot_\theta \, rot_\phi = rot_{\theta+\phi} \in H$.

Finally for each $\theta \in \mathbb{R}$, we have $rot_\theta \, rot_{-\theta} = rot_0$ so that $rot_{-\theta}$ is the inverse of rot_θ; and if $rot_\theta \in H$, $rot_{-\theta}$ also belongs to H.

Thus $H < O_2$.

(*Note :* This particular subgroup of O_2 is called *special orthogonal group* and we have the notation SO_2 for it.)

Problem 3.13 : Show that the set of all integral powers of $a \in G$, *i.e.*, $H = \{a^n : n \in \mathbb{Z}\}$ is a subgroup of G, where (G, *) is any group.

Solution : For any $r, s, \in \mathbb{Z}$, $a^r a^s = a^{r+s} \in H$, where $a^r, a^s \in H$. (*Hint :* For intgral Power of *a*, see 4.1)

Review Exercise on Chapter 3

1. Prepare group table for the group $\mathbb{Z}_6$, find all its subgroups and make lattice diagram for $\mathbb{Z}_6$.

2. Apply one of the subgroup criterias to check if
   (i) the subsets $H = \{\rho_0, \rho_1, \mu_1\}$ and $K = \{\rho_0, \mu_3\}$ of S_3 are subgroups of S_3, and
   (ii) the subsets $H = \{\rho_0, v_2, \tau_2\}$ and $K = \{\rho_0, v_1, v_2, v_3\}$ are subgroups of D_4.

3. Let $H = \{\theta \in \mathbb{R} : \sin \theta = 0\}$ $K = \{\Theta \in \mathbb{R} : \cos \theta = 0\}$. Show that $H < (\mathbb{R}, +)$ but $K \not< (\mathbb{R}, +)$.

4. Let H be set of all those permutations in

the group S_4 which carries elements 4 to 4 itself. Show that $H < S_4$.

5. Show that the set of nth roots of unity is a subgroup of the group $(\mathbb{C}^*, \cdot)$

6. Let G be an abelian group and
$H = \{a \in G : a^2 = e\}$. Show that $H < G$.

7. Let $G \times G$ be the direct product of a group G and $H = \{(a, a) : a \in G\}$.
Show that $H < G \times G$.

8. Show the following :

(i) the group $(\mathbb{Z}, +)$ acts on the set of all rational numbers Q by the product $*$ defined by

$q * n = q + n$, where $q \in \overset{*}{Q}$ and $n \in \mathbb{Z}$

(ii) the group $(\mathbb{R}^*, \cdot)$ acts on the set $\mathbb{R}^2 = \{(a, b) : a, b \in \mathbb{R}\}$ by the product $*$ defined by

$(x, y) * c = (cx, cy)$, where $c \in \mathbb{R}^*$ and $(x, y) \in \mathbb{R}^2$.

Find also, orb (1, 1) and orb (0, 0).

4 CYCLIC GROUPS

4.1 General Integral Powers

Let $(G, *)$ be a group, $a \in G$ and m be a positive integer. By associative law, $a * (a * a) = (a * a) * a$ so that the resulting value of $a * a * a$ is independent of the manner of grouping. Similarly, in view of generalised associative law the value of resulting operation $a * a * \ldots * a$ (m times) is independent of the manner of grouping (parenthesising), hence the value will be unique, any a^m. We thus define :

Definition 4.1.1 : (i) $a^m = a * a * \ldots * a$ (m times), where $m \in \mathbb{Z}^+$

For negative integers $-m$, we define
$$a^{-m} = (a^m)^{-1}$$
But, by (i), $(a^m)^{-1} = (a * a * \ldots * a)^{-1}$,
$\qquad = (a^{-1} * a^{-1} * \ldots * a^{-1})$ (by generalised form of the result $(a * b)^{-1} = b^{-1} * a^{-1}$ where a, $b \in G$)
$\qquad = (a^{-1})^m$,

so that for negative integral values, the definition is
(ii) $\qquad a^{-m} = (a^m)^{-1} = (a^{-1})^m$, $a \in G$ and $m \in \mathbb{Z}^+$

Finally, to complete the meaning of the symbol a^n, where $a \in (G, *)$, for *any* integer n, we define

(iii) $a^0 = e$, where e is the id. element of G.

Thus from (i), (ii) and (iii) above, a^n is defined for any integer n, positive, negative or zero.

We can combine two such integral powers by the following rules :

Definition 4.1.2 : (iv) $a^r * a^s = a^{r+s}$, and

(v) $(a^r)^s = a^{rs}$, for any integers r and s and for any operation $*$.

In particular, when $* = +$, definitions (i) – (v) above take the forms :

Definition 4.1.3 : (vi) $a^m = a + a + \ldots + a$ (m times), $m \in \mathbb{Z}^+$

$$= ma$$

(vii) $\quad a^{-m} = (a^m)^{-1} = (a + a + \ldots + a)^{-1}$ (m times in the bracket)

$$= a^{-1} + a^{-1} + \ldots + a^{-1}$$
$$= ma^{-1}$$
$$= m(-a) \quad \text{(when } * = +, a^{-1} = -a\text{)}$$
$$= -ma,$$

(viii) $\quad a^0 = o$, the id, element for $*$,

(ix) $\quad a^r * a^s = ra + sa$
$$= (r + s)a, \text{ and}$$

(x) $\quad (a^r)^s = s(ra)$
$$= (sr)a.$$

Examples

(1) In the group $(\mathbb{Z}, +)$,

$$3^4 = 3 + 3 + 3 + 3 \text{ (4 times)}$$
$$= 12, \text{ and}$$
$$3^{-4} = (3^4)^{-1}$$
$$= \text{Inverse of the elements } 3^4 \text{ or the element } 12$$
$$= -12.$$

When $\quad * = +, e = 0$ so that
$$3^0 = 0$$

Cyclic Groups

(2) From the adjoining group table of $(\mathbb{Z}_4, +_4)$,

$3^5 = 3 * 3 * 3 * 3 * 3$ (5 times)

$ = 2 * 2 * 3$

$ = 0 * 3$

$ = 3$

$3^{-5} = (3^5)^{-1}$

$\phantom{3^{-5}} = $ Inverse of the element 3^5, i.e., 3

$\phantom{3^{-5}} = 1$

$* = +_4$	0	1	2	3
0	0	1	2	3
1	1	2	3	0
2	2	3	0	1
3	3	0	1	2

In this example, how much is 3^0 ?

4.2 Cyclic Groups

We have seen in 4.1 that for any integer n we have a value for a^n. Let G be a group and $a \in G$. If a^n gives us all the elements of G, for some integer n, then we say that the element a has generated the entire set G and that G is a cyclic **group** generated by its element a.

Definition 4.2.1 : If in a group $(G, *)$ all the elements can be expressed as a^n, for some value of integer n, where $a \in G$ then the group $(G, *)$ is called a *cyclic group* and the element a is called *generator* of the cyclic group.

A cyclic group $(G, *)$ generated by its element a is denoted by writing $G = <a>$. In general, a cyclic group has atleast two **generators**.

This elements of a cyclic group $G = <a>$ will be of the forms

$$\ldots a^{-3}, a^{-2}, a^{-1}, a^0, a^1, a^2, a^3, \ldots , \qquad \ldots(1)$$

where $a^0 = e$. Notice that the elements in (1) may or may not all be different, depending on which we have two types of cyclic groups.

4.2.1 Infinite and finite cyclic groups

(i) Let the elements in (1) above are all distinct, i.e., for an two different integers r and s, $a^r \neq a^s$ Then there will be infinite number of elements in (1) and consequently the cyclic group $<a>$ will be an *infinite cyclic group*.

Notice that $a^r \neq a^s$ for $r \neq s$, in view of definition of order of an element, implies that $O(a) = \infty$ so that when G is an infinite cyclic group <a>, order of its generator $a = \infty$.

(ii) If the elements in (1) are not all distinct, i.e., for any two different integers r and s if $a^r = a^s$ then there will be finite number of elements in (1) and consequently the cyclic group <a> will be a *finite cyclic group*.

Let $r > s$ then $\qquad a^r = a^s$

$\Rightarrow \qquad\qquad\qquad a^r * a^{-s} = a^s * a^{-s}$

$\Rightarrow \qquad\qquad\qquad a^{r-s} = a^0 = e$

$\Rightarrow \qquad\qquad\qquad a^p = e$, (taking the integer $r-s$ as p)

Thus when cyclic group <a> is finite, there exist an integer p such that $a^p = e$ and hence order of its generator a will be finite in this case.

4.2.2 Illustrative Examples

(1) In any group (G, *) the improper subgroup {e} is cyclic because {e} = <e>.

(2) All elements of group of integers ($\mathbb{Z}$, +) are expressible as some integral powers of its elements 1 and − 1 as follows :

Elements of ($\mathbb{Z}$, +) as integral powers of 1 : $1^0 = e = 0$
$1^1 = 1$, $1^2 = 1 * 1 = 1 + 1 = 2$, $1^3 = 1 * 1 * 1 = 1 + 1 + 1 = 3$, ...,

$1^m = m$ (m a positive integer), and

$1^{-1} = (1^1)^{-1}$ = inverse of $1 = -1$. $1^{-2} = (1^2)^{-1}$

= Inverse of $2 = -2, ..., 1^{-m} = (1^m)^{-1}$ = Inverse of $m = -m$

Thus the element 1 generates all the elements of $\mathbb{Z}$, hence $\mathbb{Z}$ = <1> and the group ($\mathbb{Z}$, +) is cyclic. Similarly you can show that $\mathbb{Z}$ = <−1>. Since the group ($\mathbb{Z}$, +) is an infinite group, the cyclic group ($\mathbb{Z}$, +) is an infinite cyclic group having two generators 1 and −1. Notice that −1 = Inverse of 1.

Cyclic Groups

(3) The infinite groups $(\mathbb{Q}, +)$, $(\mathbb{Q}^*, \cdot)$, $(\mathbb{R}, +)$, $(\mathbb{R}^*, \cdot)$, $(\mathbb{C}, +)$ and $(\mathbb{C}^*, \cdot)$ are all non-cyclic because there are no generators for them.

(4) Consider the finite group $G = \{-1, 1, i, -i\}$ ($i^2 = -1$) of fourth roots of unity where $* = \cdot$. Its element i generates all the elements of G because $i^0 = e = 1$, $i^1 = i$, $i^2 = i * i = i \cdot i = -1$, and $i^3 = i * i * i = i \cdot i \cdot i = i^2 i = -i$.

Thus $G = <i>$ so that G is a finite cyclic group. Similarly $-i$ also generates the entire set G, so $G = <-i>$.

In general the group of nth roots of unity where $* = \cdot$ is a finite cyclic group generated by the elements $e^{2\pi i/n} = \cos\left(\dfrac{2\pi}{n}\right) + i \sin\left(\dfrac{2\pi}{n}\right)$.

(5) For Klein's 4 group $K_4 = \{e, a, b, c\}$, we have (See 1.5.1 (3))

$a^0 = e$, $a^1 = a$, $a^2 = a * a = e$, $= a^3 = a * a * a = a^2 * a = e * a = a$, $a^4 = a * a * a * a = e$ and similarly for other elements b and c, so that no element of set K_4 is able to generate the set K_4. Thus Klein's 4 group is not a cyclic group.

(6) From the group table of the group $(\mathbb{Z}_4, +_4)$ we have (refer the table given in example 2, §4.1), $3^0 = e = 0$, $3^1 = 3$, $3^2 = 3 * 3 = 3 +_4 3 = 2$, $3^3 = 3 +_4 3 +_4 3 = 2 +_4 3 = 1$, so that $\mathbb{Z}_4 = <3>$. Similarly $\mathbb{Z}_4 = <1>$. Thus $(\mathbb{Z}_4, +_4)$ is a finite cyclic group having generators 1 and 3.

You can verify by drawing their group-tables that all $\mathbb{Z}_n$-groups $(\mathbb{Z}_n, +_n)$ where $\mathbb{Z}n = \{0, 1, 2, ..., n\}$ are cyclic their two generators being 1 and $(n-1)$.

(7) The group $(\{1, 2, 3, 4, 5, 6\}, * = X_7)$ is cyclic generated by 3, but the group $(\{1, 3, 5, 7\}, X_8)$ is not cyclic as it has no generator.

(8) The rotation group of a regular pentagon, where the set $G = \{e, \text{rot}_{2\pi/5}, \text{rot}_{4\pi/5}, \text{rot}_{6\pi/5}, \text{rot}_{8\pi/5}\}$, is a cyclic group with generators $r = \text{rot } r_{\pi/5}, r^2, r^3$ and r^4. Similarly

the rotation group of a regular hexagon is cyclic, generators being $rot_{2\pi/6}$ and $rot_{10\pi}/6$. In general, $\forall\ n \in \mathbb{Z}^+$ the rotation group of the regular n – gon is cyclic with element $rot_{2\pi/n}$ as its generator.

Definition 4.2.2. : If two generators a and b of a cyclic group G satisfy following set of four relations

$a^4 = e$, $a^2 = b^2$, $b^4 = e$ and $a^3 b = ba$ then G is called a quaternion group.

Examples

The cyclic group of 2×2 matric with their elements from $\mathbb{C}$ with group operation $*$ as product of two such matrices, has generators $A = \begin{pmatrix} 0 & 1 \\ -1 & 0 \end{pmatrix}$ and $B = \begin{pmatrix} 0 & i \\ i & 0 \end{pmatrix}$ and is a quaternion group.

This group is a subgroup of the general linear group GL $(2, e)$.

4.2.3 Number of generators of a cyclic group

It will be seen that an infinite cyclic group has only two generators. For number of generators of a finite cyclic group we have the result in terms of Euler's function $\phi(n)$ which is defined as follows :

Definition 4.2.3 : The symbol $\phi(n)$, called Euler's function, denotes the total number of positive integers which are prime to n and leas than n. Thus if p is a prime number =, $\phi(p) = p - 1$, and in general, $\phi(n)$ = order $\{m \in \mathbb{Z}^+ : O < m < n$ and $g\ cd\ (m, n) = 1.\}$

Examples

(1) If $n = 6$ then $\phi(n) = 2$ because the positive integers 1 and 5 (both are less than 6) are prime to 6. Thus $\phi(6) = 2$. Similarly $\phi(12) = 4$.

(2) If $n = 7$ then since positive integers 1, 2, 3, 4, 5 and 6 are all less than 7 and are prime to 7 so $\phi(7) = 6$.

Cyclic Groups

Number of generators of a finite cyclic group : If $G = <a>$ is a finite cyclic group with n elements then total number of generators of G will be $\phi(n)$, where $\phi(n)$ is Euler's function. Thus if $G = <a>$ is a cyclic group with 6 elements, then since $\phi(6) = 2$, G will have two generators namely a^1 and a^5 (See example 1 above and if $G = <a>$ is a cyclic group with 7 elements, then since $\phi(7) = 6$, G will have six generators namely a^1, a^2, a^3, a^4, a^5 and a^6 (see example 2 above).

This result is contained in a more general result that we have proved in Theorem 4.3.7. Thus, regarding total number of generators possessed by a finite cyclic group, we have :

(i) If $G = <a>$ is a cyclic group of order n then $G = <a^r>$ where r is a positive integer less than n and prime to n. In other words, if a is a generator of a cyclic group of order n then other generators of G will be of the forms a^r where $0 < r < n$ and gcd (*hcf*) of r and $n = 1$.

(ii) If G is a cyclic group whose order is a prime number p then all the elements of G except id. element e will be generators of G.

4.3 Properties of cyclic groups

In this section we have arranged the common properties possessed by cyclic groups.

Theorem 4.2.1 : If G is a cyclic group generated by the element a then G is also generated by the inverse of a.

Proof : Take an arbitrary element b of group G, where $G = <a>$ is a cyclic group. Then the element b can be written as some integral power of a. Let for some integer n, $b = a^n$. But $a^n = (a^1)^{-n}$, by definition of general integral powers of a, so that $b = (a^{-1})^{-n}$, where $-n$ is some integer. Since b is arbitrary, $b = (a^{-1})^{-n}$ implies that $G = <a^{-1}>$. Thus whenever $G = <a>$ then $G = <a^{-1}>$.

Theorem 4.3.2 : (*a*) Every cyclic group is abelian.

(*b*) An abelian group may not be cyclic.

Proof for (a) : Take two arbitrary elements x and y of group $(G, *)$ where $G = <a>$ is a cyclic group. Then for some integers r and s, we have $x = a^r$ and $y = a^s$.

Then $x * y$
$$= a^r * a^s$$
$$= a^{r+s} = a^{s+r} = a^s * a^r = y * x,$$
$$(r + s = s + r, \text{ as } r, s \in \mathbb{Z})$$

showing that arbitrary elements x and y of G, satisfy commutative law. Hence G is an abelian group. Thus whenever G is cyclic, G is abelian.

Examples for (b) : Klein's group K_4 is a finite abelian group but it is not cyclic. Similarly group $(\mathbb{R}, +)$ is an infinite abelian group but it is not cyclic. Thus an abelian group need not be cyclic.

(*Note* : The group $(\mathbb{Z}_n, +_n)$, the group of nth roots of unity and the group of integers $(\mathbb{Z}, +)$ are abelian as well cyclic).

Corollary : If G is non-abelian then G is non-cyclic.

Theorem 4.3.3 : The order of a cyclic group is same as order of any of its generators, *i.e.*, if $G = <a>$ is a cyclic group then $O(G) = O(a)$.

Proof : Since $G = <a>$, $G = \{... a^{-3}, a^{-2}, a^{-1}, a^0 = e, a^1, a^2, a^3 ...\}$

Case 1: If $O(a) = \infty$ then no two integral powers of a will be equal, for if $a^p = a^q$ for some integers p and q such that $p > q$ then $a^{p-q} = e$ which is not possible as $O(a) = \infty$. Hence all the elements of G, given above, are distinct so that $O(G) = \infty$. Thus $O(a) = \infty \Rightarrow O(G) = \infty$, or $O(a) = O(G)$.

Case 2 : If $O(a) = m$, $m > 0$ then we have to prove that $O(G) = m$. Now $O(a) = m \Rightarrow a^m = e$. We will show that G has following m distinct elements

$$a^1, a^2, a^3, ..., a^q, a^p, ..., a^m = e \qquad ...(1)$$

Cyclic Groups

so that $O(G)$ will be m.

(i) Firstly, notice that all the elements in (1) are distinct, for if $a^p = a^q$ where $1 \leq q < p < m$ then $a^p * a^{-2} = a^q * a^{-q}$ or $a^{p-q} = e$,

which is not possible as $O(a) = m$ and $p-q < m$. So, no two elements in (1) can be equal. (ii) Secondly, if for some integer l, let a^l be an element of G Then for some integers q and r and for positive integer m we have by division algorithm

$$l = mq + r, 0 \leq r < m$$

so that
$$\begin{aligned} a^l &= a^{mq+r} \\ &= a^{mq} * a^r \\ &= (a^m)^q * a^r \\ &= e * a^r \end{aligned}$$

or $a^l = a^r$, showing that the element a^l is one of the m-elements given in (1). Thus by (i) and (ii), G has exactly m distinct elements given in (1), i.e., $O(G) = m$.

Examples on Theorem 4.3.3 : (i) In $(\mathbb{Z}, +)$, $\mathbb{Z} = <1>$ and $\mathbb{Z} = <-1>$. $O(\mathbb{Z}) = \infty$ and orders of elements 1 and -1 are also ∞.

(ii) In $(\mathbb{Z}_5, +_5)$, $\mathbb{Z}_5 = <1>$ and $\mathbb{Z}_5 = <4>$. $O(\mathbb{Z}_5) = 5$ and notice form the group table of $\mathbb{Z}_5$ that $O(1) = O(4) = 5$.

Corollary : If $G = <a>$ is a finite cyclic group of order m, then $G = \{a^1, a^2, a^3, ..., a^{n-1}, a^m = e\}$

Theorem 4.3.4 : An infinite cyclic group has only two generators.

Proof : Let $G = <a>$ be an infinite cyclic group. Since $G = <a>$, we know by Theorem 4.3.1 that $G = <a^{-1}>$. We will show that a and a^{-1} are the only two generators of G.

By previous theorem, $O(G) = O(a)$ and given that $O(G) = \infty$, so $O(a) = \infty$. Elements having order ∞ are also said to have order O, thus

$$a^0 = e \qquad \qquad ...(1)$$

Let a^l be another generator of G, then for same integer n, $a = (a^l)^n$

$$\Rightarrow \quad e = a^{ln} * a^{-1}$$
$$= a^{ln-1} \quad \ldots(2)$$

By (1) and (2), $\quad ln - 1 = 0$

$\Rightarrow ln = 1$, which provides two possibilities. Either $l = n = 1$ or $l = n = -1$ so that the integr l can take only two values 1 and -1, i.e., only generators of G are a and a^{-1}.

Examples on Theorem 4.3.4 : The group $(\mathbb{Z} +)$ and $(2\mathbb{Z} +)$ both are infinite cyclic groups. Their only generators are respectively 1, -1 and 2, -2. Similarly the group $(\{6^n : n \in \mathbb{Z}\};)$ is an infinite cyclic group, its only two generators being 6 and 1/6.

Theorem 4.3.5 : (a) A subgroup of a cyclic group is also cyclic.

(b) Set an example to show that a non cyclic group may have cyclic subgroups.

Proof for (a): Let $G = \langle a \rangle$ be a cyclic group and H be a subgroup of G. For improper subgroup $\{e\}$ the result is obvious as $\{e\} = \langle e \rangle$. For proper subgroup H, let m be the least positive integer such that $a^m \in$ H (This is always possible because elements of H are of the forms a^n for some integer n). We will prove that such an element a^m generates H. Take an arbitrary element b of H. Then for some integer s $b = a^s$. By division algorithm, we have integers q and r and positive integer m satisfying $s = mq + r$, where remainder r satisfies $0 \leq r < m$. Thus

$$a^s = a^{mq+r}$$
$$= a^{mq} * a^r$$
$$\Rightarrow (a^{mq})^{-1} * a^s = (a^{mq})^{-1} * a^{mq} * a^r$$
$$= e * a^r$$
$$= a^r$$

Cyclic Groups

Since $a^m \in H$ so $a^{mq} \in H$ as $mq > 0$ (In case $mq < 0$ say $mq = -8$ then $a^{mq} = a^{-8} = (a^8)^{-1} \in H$ as $H < G$), $a^s \in H$ thus left member $(a_{mq})^{-1} * a^s \in H$ or, $a^r \in H$. But $r < m$ so $a^r \in H$ is possible only when $r = 0$, giving us $s = mq$ or $a^s = a^{mq}$

$\Rightarrow b = (a^m)^q$. Since b is arbitrary element of H, H = $<a^m>$ or H is cyclic.

Example for (b) : Klein's group K_4 is a non-cyclic group but all of its subgroups namely $\{e, a\}$, $\{e, b\}$ and $\{e, c\}$ are cyclic, their generators being respectively a, b and c.

Note : We have seen in above theorem that if G = $<a>$ and H is a proper subgroup of G such that $a^m \in H$ where m is smallest positive integral power then H = $<a^m>$. Such situation can be illustrated by the cyclic group $(\mathbb{Z}, +)$ and its subgroup $(2\mathbb{Z}, +) \cdot \mathbb{Z} = <1>$ and the smallest positive integral power of 1 belonging to $2\mathbb{Z}$ is 1^2 which is $1 * 1 = 1 + 1 = 2$ and $2\mathbb{Z} = <1^2>$ or $2\mathbb{Z} = <2>$.

Theorem 4.3.6. A subgroup of an infinite cyclic group is also infinite.

Proof : Let G = $<a>$ be an infinite cyclic group and H is its subgroup. If m is the smallest positive integer such that $a^m \in H$, then we know hat H = $<a^m>$. We have to prove that H is infinite. Assume the contrary and let H has j elements so that $O(a^m) = j$

or $\qquad (a^m)^j = e$

$\Rightarrow a^{mj} = e$, which implies that $O(a)$ is finite as mj is finite. Since $O(G) = O(a)$ we arrive at the contradiction that $O(G)$ is finite.

Combining theorems 4.3.5 and 4.3.6 we see that subgroup of an infinite cyclic group is an infinite cyclic group.

Corollary : An infinite cyclic group has infinitely many subgroups, each itself being an infinite cyclic group.

The corollary follows immediately from Theorem 4.3.6 if in the proof we replace m by any other positive integer, say m_1.

Theorem 4.3.7 : If a cyclic group G has n elements and G = <a> then G = <a^p> if and only if n and p are relatively prime to each other.

Proof : (i) *First part* : Let n and p be relatively prime to each other so that for integers r and s we have $rp + sn = 1$. Let H = <a^p> be the cyclic subgroup of G then H $\subsetneq$ G.

$$rp + sn = 1 \text{ implies } a^{rp+sn} = a^1$$
$$\Rightarrow \quad a^{rp} * a^{sn} = a$$
$$\Rightarrow \quad a^{rp} = a \text{ (since } a^{sn} = (a^n)^s = e, \text{ by corollary of Theorem 4.3.3)}$$
$$\Rightarrow \quad (a^p)^r = a$$
$$\Rightarrow \quad G \subseteq H$$

so that $\quad G = H = \langle a^p \rangle$.

(ii) *Second part* : Let G <a^p> and assume the contrary that n and p are not relative primes so that if d = greatest common divisor of n and p then $d > 1$. Then $\frac{n}{d}$ and $\frac{p}{d}$ are integers, and

$$(a^p)^{n/d} = (a^n)^{p/d} = e^{p//d} = e, (a^n = e)$$

showing that $o(a^p) \leq n/d < n$, which is a contradiction as $O(a^p) = O(G) = n$. Thus gcd of n and $p = 1$, or n and p are relatively primes.

Theorem 4.3.8 : An infinite cyclic group is isomorphic to the group of integers ($\mathbb{Z}$, +).

Proof : Let G = <a> be an infinite cyclic group so that G = $\{a^n : n \in \mathbb{Z}\}$

Cyclic Groups 163

$$= \{\ldots a^{-3}, a^{-2}, a^{-1}, a^0 = e, a^1, a^2\, a^3, \ldots\}$$

Define a map $\sigma : G \to \mathbb{Z}$ by the rule $a^n \sigma = n$, for some integer n and for all $a \in G$.

Then

(i) $\{a^{n_1} * a^{n_2}\} \sigma = a^{n_1+n_2} \sigma,\ a^{n_1}, a^{n_2} \in G$ and $n_1, n_2 \in \mathbb{Z}$

$\qquad = n_1 + n_2$

$\qquad = a^{n_1} \sigma + a^{n_2} \sigma$, showing that

σ is a homomorphism

(ii) It is easy to check that σ is one-to-one and onto map. Thus σ is an isomorphism, and $G \cong (\mathbb{Z}, +)$ where $G = \langle a \rangle$ is an infinite cyclic group.

Note : In a cyclic group $G = \langle a \rangle$, since $o(a) = \infty \Rightarrow o(G) = \infty$, we can restate above theorem as :

"A cyclic group $G = \langle a \rangle$ is isomorphic to the group of integers if and only if $O(a) = \infty$".

Theorem 4.3.9 : A finite cyclic group of order m is isomorphic to the group of residue classes modulo m with operation $* = +_m$.

Proof: Let $G = \langle a \rangle$ be a cyclic group of order m so that $G = \{a^1, a^2, \ldots, a^m = e\}$ and let G' be the group of residue classes of modulo m with $* = $ addition modulo $+_m$ so that $G' = \{\bar{0}, \bar{1}, \bar{2}, \ldots, \overline{m-1}\}$. Define a map $\sigma : G \to G'$ by the rule $a^s \sigma = 1 \le \bar{s}$, $s \le m$ for all $a^s \in G$. (i) For integers s_1, s_2 $a^{s_1} = a^{s_2} \Rightarrow a^{s_1-s_2} = e$ $\Rightarrow m$ divides $s_1 - s_2$ (since $o(a) = m$) $\Rightarrow s_1 \equiv s_2 \pmod{m} \Rightarrow \bar{s}_1 = \bar{s}_2 \Rightarrow a^{s_1} \sigma = a^{s_2} \sigma$. Thus $a^{s_1} = a^{s_2} \Rightarrow a^{s_1} \sigma = a^{s_2} \sigma$, or the map σ is well-defined. (ii) If we revert the above steps, we get : $\bar{s}_1 = \bar{s}_2 \Rightarrow a^{s_1} = a^{s_2}$ or that σ is one-to-one map. (iii) For any $\bar{s} \in G'$, we have integer s such that $a^s \in G$ or that σ is onto. (iv) Finally $(a^{s_1} a^{s_2}) \sigma = (a^{s_1+s_2}) \sigma = \overline{s_1+s_2} = \bar{s}_1 + \bar{s}_2 = a^{s_1} \sigma$

$+ a^{s_2} \sigma$ or that σ is a homomorphism. From (i) to (iv), we see that the map σ is an isomorphism and hence $G \approx G'$.

Note : (1) In a cyclic group $G = <a>$, since $O(a) = m \Rightarrow O(G) = m$, we can restate above theorem as : "A cyclic group $G = <a>$ is isomorphic to the group of residue classes modulo m ($* = +_m$) if and only if $O(a) = m$".

(2) If G_1 and G_2 are two cyclic groups such that $O(G1) \neq O(G_2)$ then they can not be in isomorphism.

Theorem 4.3.10 : A finite cyclic group of order n is isomorphic to the group of nth roots of unity with $* = $.

Proof : Let $G = <a>$ be a cyclic group of order n so that $O(a) = n$ and $G = \{a^1, a^2, ..., a^n = e\}$ and let G' be the group of nth roots of units ($* = \cdot$). So that $G' = \{e^{2\pi i r/n} : r = 0, 1, ... n - 1\}$. Define a map $\sigma : G \to G'$ by the rule $a^s \sigma = e^{2\pi i s/n}, 0 < S < n-1$ for all $a^s \in G$. Then the map σ can easily be shown an isomorphism and hence $G \approx G'$.

Theorem 4.3.11 : An isomorphic *image* of a cyclic group is also cyclic.

Proof : Let $G = <a>$ be a cyclic group, G^1 be any group and the map $\sigma: G \to G^1$ be an isomorphism, so that G^1 is isomorphic image of G.

For some integer n, $a^n \in G$ so that $a^n \sigma \in G^1$ and $a^n \sigma = (a * a * ... n \text{ times}) \sigma$

$$= (a\sigma) * (a\sigma) * ... * (a\sigma) \; n \text{ times}$$

$$= (a\sigma)^n, \text{ where } a\sigma \in G^1$$

(If n is negative integer say $n = -3$ then $a^{-3} \sigma = \{(a^3)^{-1}\} \sigma = \{(a * a * a)^{-1}\} \sigma = (a^{-1} * a^{-1} * a^{-1}) \sigma = (a^{-1} \sigma) * (a^{-1} \sigma) * (a^{-1} \sigma) = (a^{-1} \sigma)^3 = (a\sigma)^{-3}$) Thus for any integer n the element $a^n \sigma$ of G^1 is expressible as some integral power of its element $a\sigma$, hence $G^1 = <a\sigma>$ or the isomorphic image G^1 is a cyclic group.

Cyclic Groups

Example on Theorem 4.3.11 : In Theorem 4.3.10, we showed that the group of nth roots of unity is isomorphic *image* of a finite cyclic group of order n. We know that the group of nth roots of unity is cyclic (with operation $* = \cdot$) generated by $e^{2\pi i/n}$. Similarly by Theorem 4.3.8, the group of integers $(\mathbb{Z}, +)$ is isomorphic image of an infinite cyclic group and $\mathbb{Z} = <1>$. Thus isomorphism preserves the property of being cyclic.

4.4 Cyclic Subgroups

We have seen in Theorem 4.3.5 that all the subgroups of a cyclic group are themselves cyclic and that non-cyclic groups may also have cyclic subgroups. Thus a wide class of subgroups are cyclic. In this section we will learn to construct cyclic subgroups of a group.

4.4.1 Cyclic subgroup of any group

We can construct cyclic subgroup of a group using the method contained in the following theorem :

Theorem 4.4.1 : Let G be a group and $a \in G$. Then the collection H of all integral powers of a, i.e., the collection H = $\{a^n : n \in \mathbb{Z}\}$ is a cyclic subgroup of G generated by its element a.

The collection H is a subgroup of G follows from Prob. 3.13, chapter 3. Since all the elements of H are in some integral powers of a, H is generated by a and hence H = $<a>$ is a cyclic subgroup of G.

Note : (1) Cyclic subgroup of a group G obtained by collecting all integral powers of its element a will be denoted by $<a>$. Thus if $b \in G$, the cyclic subgroup $<b>$ of G will be given by $<b>$ = $\{... b^{-3}, b^{-2}, b^{-1}, b^0 = e, b^1, b^2, b^3, ...\}$

(2) By Theorem 4.3.3, if the element a is of infinite order then order of cyclic subgroup $<a>$ will be infinite and if the element a is of finite order then order of cyclic subgroup $<a>$ will be finite and equal to the order of a.

(3) The cyclic subgroup $<a>$. generated by element a of a group G is the smallest subgroup of G containing a. Thus if $H < G$ such that $a \in H$ then $<a> \subseteq H$.

Examples on Theorem 4.4.1 :

(1) Consider non-cyclic group $(\mathbb{R}, +)$ and its element, say 3. Collect all integral powers of 3, namely ... $3^{-3}, 3^{-2}, 3^{-1}, 3^0 = 0, 3^1, 3^2, 3^3, ...$, which in view of operation + are ... $-9, -6, -3, 0, 3, 6, 9, ...$ so that if $H = \{... -9, -6, -3, 0, 3, 6, 9, ...\}$ then H forms a subgroup of $(\mathbb{R}, +)$. Notice the $H = 3\mathbb{Z}$, and $O(H) = 0(3) = \infty$. In general the set $n\mathbb{Z}$ forms a cyclic subgroup of $(\mathbb{R}, +)$ or even of $(\mathbb{Z}, +)$, where n is any integer. Similarly a cyclic subgroup of the group $(\mathbb{R}^*, \cdot)$ generated by its element 2 will be the set of all integral powers of 2, *viz*, the set $<2> = \{2^n : n \in \mathbb{Z}\}$
$= \{... 1/4, 1/2. 1, 2, 4, ...\}$

(2) Consider the group $S_3 = \{\rho_0, \rho_1, \rho_2, \mu_1, \mu_2, \mu_3\}$ which is a finite non-cyclic group. By collecting all integral powers of ρ_1 and of ρ_2, in both cases, we get $\{\rho_1, \rho_2, \rho_0\}$ which is the alternating group A_3. Thus

$<\rho_1> = A_3$, $<\rho_2> = A_3$ or A_3 is a cyclic subgroup of S_3. Notice that $O(A_3) = O(\rho_1) = 3$ and $O(A_3) = O(\rho_2) = 3$. Similarly $<\mu_1> = \{... \mu_1^{-2}, \mu_1^{-1}, \mu_1^0, \mu_1^1, \mu_1^2, ...\} = \{\rho_0, \mu_1\}$, $<\mu_2> = \{\rho_0, \mu_2\}$ and $<\mu_3> = \{\rho_0, \mu_3\}$, all of which are subgroups of S_3. Thus be collecting all integral powers of μ_1, μ_2, μ_3 we get respectively cyclic subgroups $<\mu_1>, <\mu_2>$ and $<\mu_3>$ of the group S_3. Notice that $A_3 = \{\rho_0, \rho_1, \rho_2\}, \{\rho_0, \mu_1\}, \{\rho_0, \mu_2\}$ and $\{\rho_0, \mu_{-3}\}$ are the only (proper) subgroups of S_3, all of which are cyclic while S_3 itself is non-cyclic group.

(3) The cyclic subgroup of S_6, generated by its element $\alpha = \begin{pmatrix} 1 & 2 & 3 & 4 & 5 & 6 \\ 2 & 3 & 4 & 5 & 6 & 1 \end{pmatrix}$ will have 6 elements as $O(a) = 6$ so that $a^6 = I$, the identity permutation. Thus
$<a> = \{a, a^2, a^3, a^4, a^5, a^6 = I\}$.

Cyclic Groups

4.4.2 Cyclic subgroups of a finite cyclic group

For obtaining the result corresponding to cyclic subgroups of a finite cyclic group, we will require the following theorems :

Theorem 4.4.2 : Let G be a group and $a \in G$ such that $O(a) = m$. Then

(i) $a^n = a^r$ where $n \in \mathbb{Z}$ and r is the remainder when n is divided by m.

(ii) $a^n = e$, where $n \in \mathbb{Z}$, if only if n is a multiple of m.

(iii) $a^n = a^l$, where $n, l \in \mathbb{Z}$, if and only if $n \equiv l \pmod{m}$, i.e, $n - l$ is a multiple of m.

Proof : (i) Since m is a positive integer and n is any integer, we have by division algorithm, integers q and r such that

$n = mq + r$, where r is the remainder when n is divided by m, i.e., $0 \leq r < m$

Thus $\quad a^n = a^{mq+r}$

$\quad\quad\quad\quad = a^{mq} * a^r$, $*$ is the b.o. of G.

$\quad\quad\quad\quad = (a^m)^q * a^r$

$\quad\quad\quad\quad = e^q * a^r$ $(O(a) = (m)$

$\quad\quad\quad\quad = a^r$

(ii) If n is a multiple of m, we have $r = 0$ when n is divided by m. Thus by (i), $a^n = a^0 = e$.

If n is not a multiple of m, we have remainder $r > 0$ when n is divided by m. Thus by (i) $a^n \neq e$.

(iii) $a^n = a^l \Rightarrow a^{n-l} = e$, which in view of (ii) is possible if and only if $n - l$ is a multiple of m, or $n \equiv l \pmod{m}$.

Theorem 4.4.3 : Let G be a group and $a \in G$ such that $O(a) = m$. Then

(i) $O(a^s) = \dfrac{l}{s}$ where $s \in \mathbb{Z}$ and $l = \text{LCM } \{m, s\}$

(ii) $O(a^s) = \dfrac{m}{d}$ where $d = \text{GCD }\{m, s\}$

(iii) In particular if s is a factor of m, $O(a^s) = \dfrac{m}{s}$

Proof : (i) By definition of order, $O(a^s) = k$, where k is least positive integer such that

$$a^{sk} = e \qquad \ldots(1)$$

For $s \in \mathbb{Z}$ and by Theorem 4.4.2 (ii)

$a^s = e$ iff s is a multiple of m.

Thus by (1) sk is a multiple of m so that sk is the LCM of m and s, say l

i.e., $\qquad sk = l$ or $k = \dfrac{l}{s}$.

(ii) If $d = \text{GCD }\{m, s\}$ then since $ms = ld$, where $l =$ LCM $\{m, s\} \Rightarrow \dfrac{l}{s} = \dfrac{m}{d}$. Hence by (i), $O(a^s) = \dfrac{l}{s} = \dfrac{m}{d}$.

(iii) If s is a factor of m, $d = \text{GCD }\{m, s] = s$ and the result follows from (ii).

The following theorem provides us the method of constructing cyclic subgroup of a finite cyclic group.

Theorem 4.4.4 : Let G be a finite cyclic group of order m and G = <a>. Then the element $b \in$ G, where $b = a^s$ for some integer s, generates a cyclic subgroup <b> of G which contains $\dfrac{m}{d}$ elements where $d = \text{GCD }\{m, s\}$.

Proof : Since G = <a> and O(G) = m thus $O(a) = m$, where $m \in \mathbb{Z}^+$. If $b \in$ G then for some integer s, we can write $b = a^s$ and by Theorem 4.4.1 the element b will generate a cyclic subgroup <b> of G where O<b> = O(b)

Cyclic Groups

$$= O(a^s)$$

$$= \frac{m}{d}, \ d = \text{GCD } \{m, s\} \ \text{(by Theorem 4.4.3 (ii))}$$

Note : By Theorem 4.4.3 (*i*), $O(a^s)$ is *also* equal to $\frac{l}{s}$, where $l = \text{LCM } \{m, s\}$ therefore

$$O<b> = \frac{l}{s} \ \text{as well.}$$

Corollary : Let G be a finite cyclic group of order m and G = <a>. Then the element $b \in G$ where $b = a^s$, $s \in \mathbb{Z}$ and s is a factor of m, generates a cyclic subgroup <b> of G which contains $\frac{m}{s}$ elements.

Proof : The Corollary follows immediately in view of Theorem 4.4.3 (*iii*).

Examples on Theorem 4.4.4 :

(1) Consider the finite cyclic group $\mathbb{Z}_{12} = \{0, 1, 2, 3, ..., 11\}$ where operation $* = +\ 12$. We know that $\mathbb{Z}_{12} = <1>$.

(*i*) In Theorem 4.4.4, let $b = 3$. Since $a = 1$ and $* = +_{12}$, we can write $3 = 1^3$, *i.e.*, in the theorem $s = 3$ (Here $b = s$) and $d = \text{GCD } \{12, 3\} = 3$. Thus cyclic subgroup generated by $b = 3$, *i.e.*, <3> will have $\frac{m}{d} = 12/3$ or 4 elements given by $3^0 = 0$, $3^1 = 3$ $3^2 = 6$ and $3^3 = 9$ and we have <3> = $\{0, 3, 6, 9\}$. (The subgroup <3> will have 4 elements follows atonce from above corollary. Similarly <2>, <4> and <6> will have respectively 6, 3 and 2 elements).

(ii) When $b = 8$, we can write $b = a^s$ as $8 = 1^8$ so that $d = \text{GCD}\{12, 8\} = 4$ and $\frac{m}{d} = 12/4$ or 3. Thus cyclic subgroup $<8>$ will have $8^0 = 0$, $8^1 = 8$, $8^2 = 8 +_{12} 8 = 4$, $8^3 = 8 +_{12} 8 +_{12} 8 = 0$, etc, and

$<8> = \{0, 4, 8\}$. Similarly
$<10> = \{0, 2, 4, 6, 8, 10\}$
$<2> = \{0, 2\ 4, 6, 8, 10\}$.
$<4> = \{0, 4, 8\}$,
$<6> = \{0, 6\}$, and
$<9> = \{0, 3, 6, 9\}$.

Notice that $<2> = <10>$, $<3> = <9>$, $<4> = <8>$.

(iii) When $b = 1, 5, 7$, or 11 (these numbers are relatively primes with $m = 12$), $d = \text{GCD}\{m, s\} = 1$ in each case and by Theorem 4.4.4 the elements 1, 5, 7 and 11 will generate cyclic subgroups of $\mathbb{Z}_{12}$ containing $\frac{m}{d} = \frac{12}{1}$ or 12 elements from $\mathbb{Z}_{12}$, i.e., $\mathbb{Z}_{12}$ itself. Thus $\mathbb{Z}_{12} = <1>$, $\mathbb{Z}_{12} = <5>$, $\mathbb{Z}_{12} = <7>$ and $\mathbb{Z}_{12} = <11>$, so that the cyclic group $\mathbb{Z}_{12}$ has 4 generators, namely 1, 5, 7 and 11. Distinct (proper) cyclic subgroups of $\mathbb{Z}_{12}$ are $<2>$, $<3>$, $<4>$ and $<6>$.

(2) The order of the group $\mathbb{Z}_{13}$ is 13, which is relatively prime with any integer. Thus by Theorem 4.4.4, $d = 1$ for any element $b \in \mathbb{Z}_{13}$ and the cyclic subgroup $<b>$ will have $\frac{m}{d}$ or 13 elements so that $\mathbb{Z}_{13}$ will not have any proper (cyclic) subgroup. What about other $\mathbb{Z}n$–groups of prime orders, say $\mathbb{Z}_{17}, \mathbb{Z}_{31}$ or $\mathbb{Z}_{97}$?

4.4.3 Illustrative Examples

(1) The cyclic subgroup of the group $(\mathbb{C}^*, \cdot)$ generated by the element $e^{2\pi i/n}$ is the group of nth roots of unity. For

Cyclic Groups

example when $n = 4$, the cyclic subgroup generated by $e^{2\pi i/4}$ is $<i> = \{i^0, i^1, i^2, i^3\} = \{1, i, -1, i\}$.

(2) The cyclic subgroup of $(\mathbb{C}^*, \cdot)$ generated by its element $1 + i$ is an infinite group as $O(1 + i) = \infty$.

(3) For infinite general linear group over $\mathbb{R}$, $GL_2(\mathbb{R})$, the cyclic subgroup generated by its element $\begin{pmatrix} 0 & 1 \\ 1 & 0 \end{pmatrix}$ is finite and has two elements, namely $\begin{pmatrix} 0 & 1 \\ 1 & 0 \end{pmatrix}$ and $\begin{pmatrix} 1 & 0 \\ 0 & 1 \end{pmatrix}$.

Thus

$$< \begin{pmatrix} 0 & 1 \\ 1 & 0 \end{pmatrix} > = \left\{ \begin{pmatrix} 0 & 1 \\ 1 & 0 \end{pmatrix}, \begin{pmatrix} 1 & 0 \\ 0 & 1 \end{pmatrix} \right\}.$$

But the cyclic subgroup generated by the element $\begin{pmatrix} 1 & 1 \\ 0 & 1 \end{pmatrix}$ of $GL_2(\mathbb{R})$ is infinite given by

$$< \begin{pmatrix} 1 & 1 \\ 0 & 1 \end{pmatrix} > = \left\{ \begin{pmatrix} 1 & n \\ 0 & 1 \end{pmatrix} : n \in \mathbb{Z} \right\}.$$

(4) Since for any reflection τ in D_n (see, for example 2.3) we have

$\tau^2 = $ id. element $= \tau^4 = \tau^6 = \ldots,$

$\tau^3 = \tau = \tau^5 = \tau^7 = \ldots,$ and

$\tau^{-1} = \tau$, thus the cyclic subgroup of dihedral group Dn generated by any reflection τ in D_n has only two elements namely id. element and τ, i.e., $<\tau> = \{e, \tau\}$.

The rotation $2\pi/n$ in D_n generates a cyclic subgroup of order n which is the rotation group of the regular n-sided polygon. Write a general element of the subgroup $<rot_{2\pi/n}>$ of D_n.

It is easy to see that when n is odd, D_n has n cyclic subgroups of second order while it has $(n + 1)$ cyclic subgroups of second order when n is an even positive integer.

4.4.4 Lattice diagrams for $(\mathbb{Z}_n, +_n)$ groups

Theorem 4.4.4 enables us to find out all the cyclic subgroups of a finite cyclic group. Let us consider some $\mathbb{Z}_n$ - groups and draw their lattice diagrams.

(i) $n = 12$. As seen in example 1 on theorem 4.4.4, the group $\mathbb{Z}_{12}$ has 4 distinct proper (cyclic) groups namely <2>, <3>, <4> and <6> and two improper subgroups namely the group $\mathbb{Z}_{12}$ and <O>. (What is relation of the numbers 2, 3, 4 and 6 with the number 12, the order of the group ?)

Notice also that <4> is a subgroup of <2> and <6> is subgroup of <2> as well as of <3> (Observe the relation of 2 with 4 and that of 2, 3 with 6.) All these subgroups can be displayed in the lattice diagram of the group $\mathbb{Z}_{12}$, which looks like a family photograph of the group $\mathbb{Z}_{12}$:

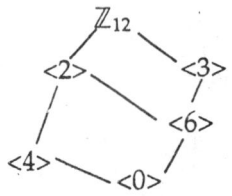

Lattice diagram for $\mathbb{Z}_{12}$

(ii) $n = 18$. The proper cyclic subgroups of $\mathbb{Z}_{18}$ are <2>, <3>, <6> and <9>.

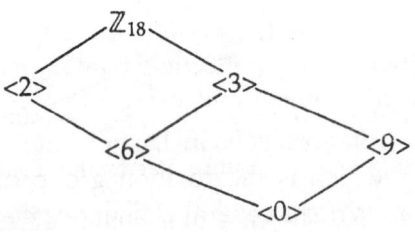

Lattice diagram for $\mathbb{Z}_{18}$

(iii) $n = 36$. Here the proper subgroups are <2>, <3>, <4>, <6>, <9>, <12> and <18>.

Cyclic Groups

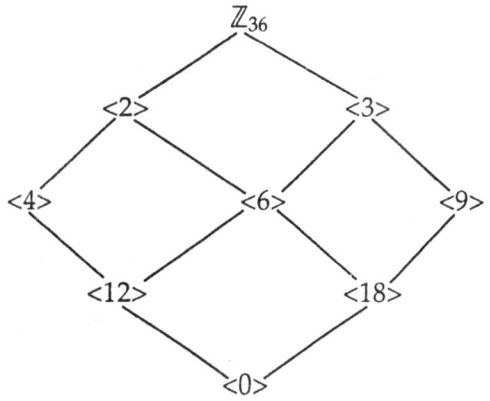

Lattice diagram for $\mathbb{Z}_{36}$

(iv) $n = 100$. The proper subgroups of $\mathbb{Z}_{100}$ are <2>, <4>, <5>, <10>, <20>, <25> and <50>.

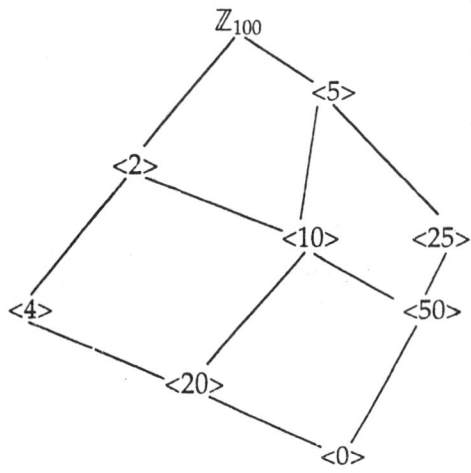

Lattice diagram for $\mathbb{Z}_{100}$

(v) $n = 60$. The case of $n = 60$ is different than those of $n = 12, 18, 36$ or 100 because, for examples, $100 = 2^2 * 5^2$; $36 = 2^2 * 3^2$, the lattice diagram of $\mathbb{Z}_{100}$ or of $\mathbb{Z}_{36}$ is two-dimensional whereas since $60 = 2^2 * 3 * 5$, the lattice diagram of $\mathbb{Z}_{60}$ and all such $\mathbb{Z}_n$-groups will be three-dimensional.

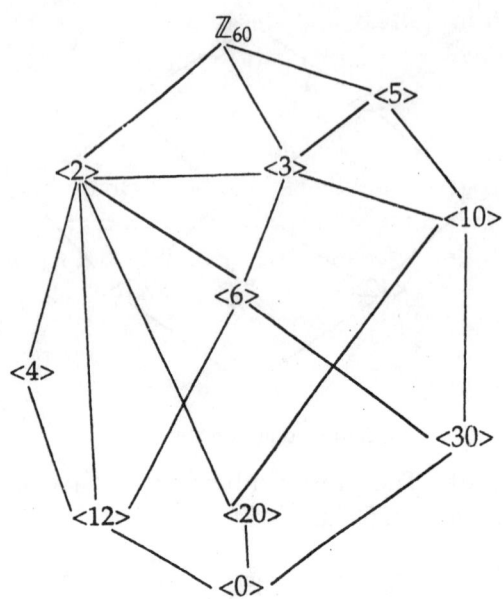

Lattice diagram for $\mathbb{Z}_{60}$

Solved Problems Set on Chapter 4

Problem 4.1 : Let $G = \langle a \rangle$ be a cyclic group such that $O(a) = \infty$. Prove that G is infinite.

Solution : When order of any elements is infinite we allot zero as its order and we write $O(a) = \infty \Rightarrow a^0 = e$. Here $G = \langle a \rangle$ and $O(a) = \infty$, we will show that no two integral powers of a can be equal. If possible, for any two integers s and t where $t > s$, let $a^t = a^s$

$\Rightarrow \qquad a^t * a^{-s} = a^s * a^{-s}$

$\Rightarrow \qquad a^{t-s} = e,$

which is not possible because order of a is infinite. Thus $a^t \neq a^s$ so that $G = \langle a \rangle$ is infinite and $G = \{... a^{-3}, a^{-2}, a^{-1}, a^0, a^1, a^2, a^3, ...\}$.

Problem 4.2 : Let $G = \langle a \rangle$ be a cyclic group such that $O(a) = m$. Prove that

$$G = \{a^1, a^2, a^3, ..., a^m = e\}$$

Cyclic Groups

Solution : Since $O(a) = m$, $a^m = e$

Firstly we will show that G has at the most following m elements

$$a^1, a^2, a^3, \ldots, a^m = e$$

If possible, let a^l, where $l > m$, also belongs to G. By division algorithm we have integers q and r such that $l = mq + r$, where the remainder r satisfies the inequality $0 \leq r < m$

Thus
$$a^l = a^{mq+r}$$
$$= a^{mq} * a^r$$
$$= (a^m)^q * a^r$$
$$= e^q * a^r$$

or, $a^l = a^r$ so that a^l, for any integer l, is one of the above $(m-1)$ listed elements of G. Secondly we will show that the m elements $a^1, a^2, a^3 \ldots, a^m = e$ are all distinct. If possible, for integers p and q where $0 < q < p \leq m$, let $a^p = a^q$

$$\Rightarrow \qquad a^{p-q} = e,$$

which is not possible as $p - q < m$ and $O(a) = m$. Thus if G = $<a>$ such that $O(a) = m$, then set G is given by

$$G = \{a^1, a^2, a^3, \ldots, a^m = e\}.$$

Problem 4.3 : If G = $<a>$ is an infinite cyclic group then order of every non-identity element in G is infinite . In other words, prove that on infinite cyclic group is torsion free.

Solution : Since $O(G) = \infty$ and G = $<a>$, $O(a) = \infty$. Take an arbitrary element g of G. Then for some integer n, $g = a^n$. If possible, let g be of finite order say m so that $g^m = e$. Then

$$g = a^n \text{ implies } g^m = (a^n)^m$$

$\Rightarrow e = a^{nm}$, where nm is some integer. But $a^{nm} = e$ is not possible as $O(a) = \infty$ thus order of every element g of group G, where $g \neq e$, is infinite, or G is torsion free.

Problem 4.4 : Let G be a finite group of order m. Prove that G is cyclic if and only if G has an element of order m.

Solution : Given that G is finite group and $O(G) = m$.

If G is cyclic and G = <a> then O(G) = O(a) = m and hence there is an element $a \in G$ whose order is also m.

Conversely let G has an element g such that O(g) = m. Then the element g generates a cyclic subgroup <g>, of G whose order is m and <g> = $\{g^1, g^2, ..., g^m = e\}$
= G (see Theorem 4.4.1 Note (2) and Cor., Theorem 4.3.3) so that G is cyclic.

Problem 4.5 : Prove that a cyclic group with only one generator can have at the most two elements.

Solution : Let G be a cyclic group generated by the elements a. Then we know that G = <a^{-1}>. Given that G has only one generator which implies that $a = a^{-1}$

$\Rightarrow \qquad a * a = a^{-1} * a$
$\qquad\qquad\quad = e$

$\Rightarrow \qquad a^2 = e$. With this condition we have two possibilities.

(i) If $a = e$ then G = {e}
(ii) If $a \neq e$ then $a^1 = a$, $a^2 = e$, $a^3 = a$, $a^4 = e$, ... and G = {e, a}. Thus G has at the most two elements.

Example of this situation is the cyclic group {–1, 1} which has only one generator, –1.

Problem 4.6 : Find the permutation group which is isomorphic to the cyclic group G = {a^1, a^2, a^3 $a^4 = e$}.

Solution : The group G has following group table

	$a^4=e$	a^1	a^2	a^3
$a^4=e$	e	a^1	a^2	a^3
a^1	a^1	a^2	a^3	e
a^2	a^2	a^3	e	a^1
a^3	a^3	e	a^1	a^2

Since O(G) = 4, we set following correspondences with the elements of the group S_4 :

Cyclic Groups

$$e \leftrightarrow \begin{pmatrix} e\ a^1\ a^2\ a^3 \\ e\ a^1\ a^2\ a^3 \end{pmatrix} = \begin{pmatrix} 4\ 1\ 2\ 3 \\ 4\ 1\ 2\ 3 \end{pmatrix} = I$$

$$a^1 \leftrightarrow \begin{pmatrix} e\ a^1\ a^2\ a^3 \\ a^1\ a^2\ a^3\ a^4=e \end{pmatrix} = \begin{pmatrix} 1\ 2\ 3\ 4 \\ 2\ 3\ 4\ 1 \end{pmatrix} = \alpha_1 \text{ (say)}$$

$$a^2 \leftrightarrow \begin{pmatrix} e\ a^1\ a^2\ a^3 \\ a^2\ a^3\ a^4=e\ a^1 \end{pmatrix} = \begin{pmatrix} 1\ 2\ 3\ 4 \\ 3\ 4\ 1\ 2 \end{pmatrix} = \alpha_2$$

and

$$a^3 \leftrightarrow \begin{pmatrix} e\ a^1\ a^2\ a^3 \\ a^3\ a^4=e\ a^1\ a^2 \end{pmatrix} = \begin{pmatrix} 1\ 2\ 3\ 4 \\ 4\ 1\ 2\ 3 \end{pmatrix} = \alpha_3$$

Thus the permutation group isomorphic to the cyclic group $G = \{a^1, a^2, a^3, a^4 = e\}$ is the group $\{I, \alpha_1, \alpha_2, \alpha_3\}$, which is a subgroup of the group S_4.

Note : The above problem illustrates the general result that every finite group of order n is of isomorphic to subgroup of a permutation group S_n on n symbols, which we will come across in chapter no. 7 (corollary 7.4.1).

Problem 4.7 :

(i) Show that 47 is a generator of the group $\mathbb{Z}_{100}$.

(ii) Find the cyclic subgroup of $\mathbb{Z}_{100}$ generated by its element 40.

(iii) Find the cyclic subgroup of $\mathbb{Z}_{100}$ generated by its element 50.

Solution : (i) By Theorem 4.4.4, the element 47 will generate a cyclic subgroup <47> of the group $\mathbb{Z}_{100}$ containing $\frac{m}{d}$ elements, where $\quad m = 100, d = \gcd\{m, s\}$

$$= \gcd\{100, 47\}\ (s = 47 \text{ as } 47 = 1^{47})$$

$$= 1$$

or $<47>$ will contain $\frac{100}{1}$ i.e., 100 elements

or $\qquad <47> = \mathbb{Z}_{100}$.

(ii) Here $\qquad d = \gcd\{100, 40\}$

$\qquad\qquad\qquad = 20 \qquad (s = 40, \text{ as } 40 = 1^{40})$

so that $\qquad \frac{m}{d} = 5$ and hence $<40>$
will contain 5 elements namely $40^0 = 0$, $40^1 = 40$, $40^2 = 80$, $40^3 = 40^2 +_{100} 40 = 20$ and $40^4 = 40^3 +_{100} 40 = 20 +_{100} 40 = 60$, or $<40> = \{0, 20, 40, 60, 80\}$.

(iii) By corollary of Theorem 4.4.4, the cyclic subgroup $<50>$ will contain $100/50$ i.e., 2 elements and they are $50^0 = 0$ and $50^1 = 50$ (Note that $50^2 = 50 +_{100} 50 = 0$, $50^3 = 50$, etc.) Thus $<50> = \{0, 50\}$.

Problem 4.8 : If p and q are prime numbers then show that the number of generators of the group $\mathbb{Z}_{pq}$ is $(p-1)(q-1)$.

Solution : By Theorem 4.4.4, a positive integer m will not generate $\mathbb{Z}_{pq}$ if and only if $m = o$ or m is divisible either by p or q, where p and q are prime numbers. Thus m will not generate $\mathbb{Z}_{pq}$ iff $m \in D$, where set $D = \{0, p, 2p, ..., (q-1)p, q, 2q, ...,(p-1)q\}$ Then $O(D) = 1 + (q-1) + (p-1) = p + q - 1$

Thus total number of generators of group
$$\mathbb{Z}_{pq} = pq - (p + q - 1) \qquad \text{(see 4.2.3)}$$
$$= (p-1)(q-1)$$

Problem 4.9 : If p is a prime number and m is a positive *integer* then show that the number of generators of the cyclic group $\mathbb{Z}_{p^m}$ is $p^{m-1}(p-1)$. Hence find the number of generators of the group $\mathbb{Z}_{16}$.

Solution : Let $\psi\,[p^k, p^{k+1})$ be the number of integers x satisfying $p^k \le x < p^{k+1}$ and that x is divisible by p. Then it is easy to verify that

ψ [O, p) = 1 and ψ [p^k, p^{k+1}) = $p^k - p^{k-1}$

so that

ψ [O, p^m) = ψ [O, p) + ψ [p, p^2) + ... + ψ [p^{m-1}, p^m)
= 1 + (p – 1) + ($p^2 - p$) + ... + ($p^{m-1} - p^{m-2}$)
= p^{m-1}

Now, generators of $\mathbb{Z}p^m = \phi(p^\mu)$, where ϕ is Euler's function.

= $p^m - \psi$ [O, p^m)
= $p^m - p^{m-1}$
= p^{m-1} (p– 1).

Generators of $\mathbb{Z}_{16}$:

Here $16 = 2^4$ so that $p = 2$ and $m = 4$.

Thus generators of $\mathbb{Z}_{16} = 2^{4-1}(2-1) = 8$ and they are 1, 3, 5, 7, 9, 11, 13 and 15.

Problem 4.10 : Let G = <a> be a finite cyclic group of order m, k be a positive integer such that $1 \le k < m$ and d = gcd $\{m, k\}$. Then prove that the subgroups of G generated by a^k and a^d are the same, i.e., <a^k> = <a^d>. Prove also that the total number of distinct subgroups of G is equal to the number of distinct positive divisors of m.

Solution : Since $k = sd$, for some integer s,

$a^k = a^{sd}$
$= a^{ds}$
$= (a^d)^s \in$ <a^d>

$\Rightarrow$ <a^k> $\subseteq$ <a^d> ...(1)

Also by Theorem 4.4.4, O <a^k> = O <a^d> = $\dfrac{m}{d}$, hence by (1), <a^k> = <a^d>. ...(2)

Further, let the positive divisors of m be the integers $h_1, h_2, ..., h_j$ and since d is a factor of m, $d = h_i$ for some i, $1 \le i \le j$. Further, <a^{h_1}>, <a^{h_2}>, ..., <a^{h_j}> are distinct subgroups of G, their orders being $\frac{m}{h_1}$, $\frac{m}{h_2}$, ..., $\frac{m}{h_j}$ respectively. By Theorem 4.3.5 (a), all those subgroups are of the form $<a^k>$ for some k, thus by (2) $<a^k> = <a^d> = <a^{h_i}>$, or $<a^k> = <a^{h_i}>$ where $1 \le i \le j$. Hence all the distinct subgroups of G are $<a^{h_1}>$, $<a^{h_2}>$, ..., $<a^{h_j}>$ which are j in number where j is the number of distinct positive divisors of m.

(Example on Prob. 4.10 : Example 1 on theorem 4.4.4 illustrates the result contained in this problem. Number of distinct positive divisors of 12 is 6, namely 1, 2, 3, 4, 6 and 12 and consequently by this result $\mathbb{Z}_{12}$ has in all six subgroups – two improper and for proper subgroups.)

Problem 4.11 : Let $G_1 = <a>$ be a cyclic group of order i and $G_2 = <b>$ be a cyclic group of order j such that $gcd\ \{i, j\} = 1$. Prove that the direct product $G_1 \times G_2$ is a cyclic group of order ij and is generated by (i, j).

Solution : Since $O(G_1) = i$ and $O(G_2) = j$, by definition of direct product, $O(G_1 \times G_2) = ij$. Also, $G_1 = <a>$ and $O(G_1) = i$ which implies that $O(a) = i$. Similarly $O(b) = j$. This $a^i = e_1$ – (1) and $b^j = e_2$ – (2), where e_1 and e_2 respectively are the id. elements of G_1 and G_2. Let $g = (a, b)$, then $g \in G_1 \times G_2$ and

$$g^{ij} = (a^{ij}, b^{ij}) = ((a^i)^j, (b^j)^i)$$
$$= (e_1, e_2)$$
$\Rightarrow \qquad o(g) \le ij \qquad \qquad ...(3)$

Let for integer l,
$$g^l = (e_1, e_2)$$
$\Rightarrow \qquad (a^l, b^l) = (e_1, e_2)$
$\Rightarrow \qquad a^l = e_1, b^l = e_2,$

which by Theorem 4.4.2 (ii) together with (1) and (2) implies that i divides l and j divides l. Since $gcd\ (i, j) = 1$, ij divides l

or $\qquad l \ge ij$

or $\qquad O(g) \ge ij \qquad \qquad ...(4)$

Thus, by (3) and (4) $O(g) = ij$,
or $\quad\quad\quad O(a, b) = ij$

Since $O(G_1 \times G_2) = O(a, b) = ij$ and $(a, b) \in G_1 \times G_2$, we conclude that $G_1 \times G_2$ is a cyclic group generated by (a, b), or $G_1 \times G_2 = <(a, b)>$ and $O(G_1 \times G_2) = ij$.

Problem 4.12 : Prove that every finite group of composite order has proper subgroups.

Solution : Let G be a finite group of composite order and $O(G) = rs$ where r and s are positive integers such that $r > 1$ and $s > 1$.

Case 1 : Let $\quad\quad G = <a>$ be cyclic so that
$$O(G) = O(a) = rs$$
$\Rightarrow \quad\quad\quad a^{rs} = e \quad\quad\quad\quad ...(1)$
$\Rightarrow \quad\quad\quad (a^s)^r = e$
$\Rightarrow$ order of element $a^s \leq r$

Let $O(a^s) < r$ and $O(a^s) = i$ Then $i < r$
$\Rightarrow \quad\quad\quad (a^s)^i = e$
$\Rightarrow \quad\quad\quad a^{si} = e \quad\quad\quad\quad ...(2)$

But $i < r \Rightarrow is < rs$ which, in view of (1) and (2), is not possible. Thus $O(a^s) = r$. The element a^s generates a cyclic subgroup $<a^s>$ of G where $O<a^s> = O(a^s) = r$. Since $1 < r < rs$, the (cyclic) subgroup $<a^s>$ of G is a proper subgroup of G.

Case 2 : Let G be a non-cyclic group. For any element $e \neq a \in G$, $O(a) < rs$. Take an arbitrary element $b \in G$. Then $2 \leq O(b) < rs$ and the cyclic subgroup of G generated by this element b, i.e., $<b>$, will be a proper subgroup of G as $O<b> = O(b)$.

Problem 4.13 : If a group G has no proper subgroups then prove that G must be a finite group of order either 1 or a prime number p.

Solution : Given G is a group with no proper subgroups, i.e., the only subgroups of G are $\{e\}$ and G itself.

To prove that G is finite and $O(G) = 1$ or a prime number p.

If possible, let G be an infinite group and $e \neq a \in G$ be an arbitrary element of G. We have two possibilities : either $O(a)$ is finite or $O(a)$ is infinite.

(i) Of $O(a)$ is finite. Then the cyclic subgroup $<a>$ with order $O(a)$, will be a proper subgroup of G, which is contrary to given statement that G has no proper subgroups. Thus, in this case, G must be a finite group.

(ii) If $O(a) = \infty$ then $O<a> = \infty$ If $G = <a>$ (G is assumed to be an infinite subgroup) then G will have proper (cyclic) subgroups. If $G \neq <a>$ then $<a>$ will be a proper subgroup of G, as $a \in G$. Thus in both the cases, $G = <a>$ to $G \neq <a>$, we arrive at the contradiction that G has no proper subgroups, hence G must be a finite group.

In problem 4.12, we have proved that a (finite) group of composite order always has a proper subgroup. Since G has no proper subgroups and is finite. Thus $O(G) = 1$ or a prime number p (non-composite numbers).

Review Exercise on Chapter 4

1. Show that the group $(6\mathbb{Z}, +)$ and $(\{6^n : n \in \mathbb{Z}\}, \cdot)$ are cyclic. Find their generators. How many generators will have each in total ?

2. (i) Show that the element 13 is a generator of group $\mathbb{Z}_{30}$.

   (ii) Find the cyclic subgroup of $\mathbb{Z}_{30}$ generated by 25.

3. (i) If $G = <a>$ and $O(G) = 9$, find all the generators of G.

   (ii) Find out all the generators of the group $\mathbb{Z}_8$.

4. Let a group G with atleast two elements has no proper subgroup. Prove that $G \approx \mathbb{Z}_p$, where p is prime.

5. Show that two cyclic groups of the same order are isomorphic.

6. Let $O(G) = pq$, where p and q are prime numbers. Verify by an example that every proper subgroup of G is cyclic.

Cyclic Groups

7. Let $G = \langle a \rangle$ be a finite cyclic group of order m. Prove that for any divisor d of m there exists a unique subgroup of G of order d.

8. (i) Find the number of different subgroups of a cyclic group of order 36.

(ii) Find a cyclic subgroup of group D_4 of order 2.

9. Let $G = \langle a \rangle$ be a cyclic group and $H < G$. Prove that whenever $a^2 \in H$ and $a \notin H$ then

(i) $a^{-3} \notin H$, and

(ii $a^7 \notin H$.

10. Let G be a group $a \in G$ and $O(a) = m$. Prove that the Cyclic subgroup generated by the element a has order m.

[Hint. $O(a) = m \Rightarrow a^m = e$ and for any integer n, by division algorithm $n = mq + r$, where $0 \leq r < m$. Thus $a^n = a^r$ and hence all integral powers of a are $a^0 = e, a^1, a^2, \ldots, a^{m-1}$ which are all distinct. Thus the set $\langle a \rangle = \{a^0 = e, a^1, a^2, \ldots, a^{m-1}\}$, which can be proven to form a group, has order m.]

5 COSETS AND LAGRANGE'S THEOREM

5.1 Partitions of a Group G

We will learn to partition a given group G by means of an equivalence relation defined over G and call each equivalence class a coset. Cosets play an important role specially in the study of finite groups and they are instrumental in establishing some powerful results on finite groups including Lagrange's theorem. Firstly we will learn to form two types of cosets.

5.1.1 The Sets aX and Xa

Let X be non-empty subset of a group $(G, *)$ and a be any element of G, not necessarily belonging to X. Pre-operate each element of X by this element a to get the set $\{a * x : \forall\, x \in X\}$. Denote the set by aX. Notice that a is a fixed element of G and it has been written in the left side of the notation aX. Similarly, post-operate each element of X by the element a to get another set $\{x * a : \forall\, x \in X\}$. Denote this set by Xa. The notation Xa has a in its right side. Thus we have formed two non-empty subsets aX and Xa of G with the help of an element a of G and a non-empty subset X of G.

If X is a finite subset of G, say $X = \{x_1, x_2, ..., x_n\}$ then $aX = \{a * x_1, a * x_2, ..., a * x_n\}$ and $Xa = \{x_1 * a, x_2 * a, ..., x_n * a\}$. In general $aX \neq Xa$; of course when $a = e$, $eX = Xe = X$, and when $X = \{e\}$ then $aX = Xa = \{a\}$. What about aG and Ga ?

We now restrict non-empty subset X of G to be a subgroup H of G and define cosets as follows :

Definition 5.1.1 : If H is a subgroup of G then the subset aH of G is called **a left coset** of H in G determined (or

Cosets and Lagrange's Theorem

generated) by the element a of G. The subset Ha is called a right coset of H in G determined by the element a. Thus

left coset $aH = \{a * h : \forall\ h \in H\}$ and

the right coset $Ha = \{h * a : \forall\ h \in H\}$

where a is a fixed element of group G.

Note : 1. For improper subgroups $\{e\}$ and G, the left cosets determined by a are respectively $a\{e\} = \{a * e\} = \{a\}$ and aG. Similarly the right cosets for them are $\{a\}$ and Ga.

2. If $* = +$, then $aH = \{a + h : h \in H\}$ and $Ha = \{h + a : h \in H\}$.

3. In the definition of aH, the element a can be replaced by any other element of G and hence a left coset of H in G will be determined by each element of G providing the total number of left cosets of H in G equal to the order of G. Similarly the total number of right cosets of H in G will be equal to 0 (G).

4. A coset can never be an empty set because, for example, in the right coset Ha, atreast $a \in Ha$, as $a = e * a, e \in H$.

5. When $a = e$, the cosets are eH and He and both are equal to H. Thus a subgroup H of a group G is itself a left coset eH as well as a right coset He of H in G. We will see that from all possible cosets xH and Hx of H in G only eH and He and any other left coset which is equal to H are the subgroups of G. Remaining cosets are only non-empty subsets of G.

6. When group G is abelian, $a * h = h * a$ and hence $aH = Ha$. Thus each left coset xH of a subgroup H of an abelian group is equal to the corresponding right coset Hx. For non-abelian groups, in general, $xH \neq Hx$ for any $x \in G$.

5.1.2 Illustrative Examples

1. Let $G = \{-1, 1, i, -i\}$ where $* = \cdot$ The proper subgroup of G is $H = \{-1, 1\}$.

   (*i*) Left cosets of H in G —

   When $a = -1$, left coset $-1H = \{-1 \cdot (-1), -1 \cdot 1\} = \{1, -1\} = H$

When $a = 1$, left coset $1H = \{1 \cdot (-1), 1 \cdot 1\} = \{-1, 1\} = H$

(Notice that in each case $a \in H$).

When $a = i$, left coset $iH = \{i \cdot (-1), i \cdot 1\} = \{-i, i\}$, and when $a = -i$, left coset $-iH = \{-i \cdot (-1), -i \cdot 1\}$

$= \{i, -i\} = iH$.

Thus total number of left coset of H in G is four, the order of G. Out of total four cosets namely $-1H$, $1H$, iH and $-iH$ only two are distinct which are $-1H$ (or $1H$) and iH (or $-iH$). The coset $-1H$ (whose value is H) is subgroup of G and the coset iH is not a subgroup of G. Notice also that union of distinct left cosets is equal to G, i.e. $(-1H) \cup (iH) = G$.

(ii) Right cosets of H in G —

By taking $a = -1, 1, i$ and $-i$ we get four right cosets of H in G which are H_{-1}, H_1, H_{-i} out of which H_{-1} (or H_1) and H_i (or H_{-i}) are distinct and their union equals G.

Since G is abelian, we have $xH = Hx$, for any $x \in G$.

2. Let us take a non-abelian finite group $S_3 = \{\rho_0, \rho_1, \rho_2, \mu_1, \mu_2, \mu_3\}$ and consider its subgroup $H = \{\rho_0, \mu_1\}$. We will compute cosets of H in S_3.

(i) Let cosets of H in S_3 —

When $a = \rho_0$, left coset $\rho_0 H = \{\rho_0 * \rho_0, \rho_0 * \mu_1\} = \{\rho_0, \mu_1\} = H$

When $a = \rho_1$, left coset $\rho_1 H = \{\rho_1 * \rho_0, \rho_1 * \mu_1\} = \{\rho_1, \mu_2\}$

When $a = \rho_2$, left coset $\rho_2 H = \{\rho_2 * \rho_0, \rho_2 * \mu_1\} = \{\rho_2, \mu_3\}$

When $a = \mu_1$, left coset $\mu_1 H = \{\mu_1 * \rho_0, \mu_1 * \mu_1\} = \{\mu_1, \rho_0\} = H$

When $a = \mu_2$, left coset $\mu_2 H = \{\mu_2 * \rho_0, \mu_2 * \mu_1\} = \{\mu_2, \rho_1\} = \rho_1 H$

When $a = \mu_3$, left coset $\mu_3 H = \{\mu_3 * \rho_0, \mu_3 * \mu_1\} = \{\mu_3, \rho_2\} = \rho_2 H$

Thus we have three distinct left cosets of H in S_3, namely $\rho_0 H, \rho_1 H$ and $\rho_2 H$. The union of distinct left cosets $\rho_0 H \cup \rho_1 H \cup \rho_2 H$ is equal to $\{\rho_0, \rho_1, \rho_2, \mu_1, \mu_2, \mu_3\}$, the group S_3.

Cosets and Lagrange's Theorem

Observe also that $\rho_0 H = H$ and $\mu_1 H = H$ where $\rho_0, \mu_1 \in H$.

(ii) Right cosets of H in S_3 —

When $a = \rho_0$, right coset $H\rho_0 = \{\rho_0 * \rho_0, \mu_1 * \rho_0\} = \{\rho_0, \mu_1\} = H$

When $a = \rho_1$, right coset $H\rho_1 = \{\rho_0 * \rho_1, \mu_1 * \rho_1\} = \{\rho_1, \mu_3\}$

When $a = \rho_2$, right coset $H\rho_2 = \{\rho_0 * \rho_2, \mu_1 * \rho_2\} = \{\rho_2, \mu_2\}$

When $a = \mu_1$, right coset $H\mu_1 = \{\rho_0 * \mu_1, \mu_1 * \mu_1\} = \{\mu_1, \rho_0\} = H$

When $a = \mu_2$, right coset $H\mu_2 = \{\rho_0 * \mu_2, \mu_1 * \mu_2\} = \{\mu_2, \rho_2\} = H\rho_2$

When $a = \mu_3$, right coset $H\mu_3 = \{\rho_0 * \mu_3, \mu_1 * \mu_3\} = \{\mu_3, \rho_1\} = H\rho_1$

Thus we have three distinct right cosets of H in S_3, namely $H\rho_0$, $H\rho_1$ and $H\rho_2$. Their union is also equal to the group S_3. The right cosets $H\rho_0$ and $H\mu_1$ both are equal to subgroup H where $\rho_0, \mu_1 \in H$. Notice that $\rho_0 H = H\rho_0$, but $\rho_1 H \neq H\rho_1$ and $\rho_2 H \neq H\rho_2$.

Before taking up more examples, we define left and right quotient sets.

Definition 5.1.2 : The set $\{aH : a \in G\}$ of all (distinct) left cosets of a subgroup H of group G is called the left quotient set of G by H, denoted by G\H.

Definition 5.1.3 : The set $\{Ha : a \in G\}$ of all (distinct) right cosets of a subgroup H of group G is called the right quotient set of G by H, denoted by G/H.

Example : In the above illustrative examples (1) and (2), when $G = \{-1, 1, i, -i\}$ and $H = \{-1, 1\}$, the left quotient set G\H = \{1H, iH\} and the right quotient set G/H = \{H1, Hi\}. When $G = S_3$ and $H = \{\rho_0, \mu_1\}$, G\H = \{ρ_0H, ρ_1H, ρ_2H\} and G/H = \{Hρ_0, Hρ_1, Hρ_2\}.

Is G\H = G/H in both the cases ? If not, what is the difference between two groups ?

Note :

1. When $* = +$, the left and right quotient sets will be given by

$$G \backslash H = \{a + H : a \in G\} \text{ and}$$
$$G/H = \{H + a : a \in G\}$$

2. Since, in general $aH \neq Ha$, the sets $G \backslash H$ and G/H are different sets and $G \backslash H \neq G/H$.

3. When group G is abelian, $aH = Ha$, $\forall a \in G$ and thus the sets $G \backslash H$ and G/H coincide and we write $G \backslash H = G/H$ and the common set is denoted by $G \mid H$. The set $G \mid H$ is called *the quotient set* of G by H.

In example (1) of Illustrative examples 5.1.2, since $1H = H1$ and $iH = Hi$, thus $G \backslash H = G/H$ and $G \mid H = \{1H, iH\}$, while in example (2), since $\rho_1 H \neq H\rho_1$ and $\rho_2 H \neq H\rho_2$, we have $G \backslash H \neq G/H$.

We will now show that the sets $G \backslash H$ and G/H are the two different partitions of G. For proving it, we require the following result :

Lemma 5.1.1 : Let $a, b \in G$ and Ha be the right coset of H determined by a. Then $b \in Ha$ iff $ba^{-1} \in H$. Similarly if aH is the left coset of H determined by a then $b \in aH$ iff $a^{-1}b \in H$.

The proof of lemma is immediate, because $b \in Ha$ iff $b = ha$ for some $h \in H$, which implies $h = ba^{-1}$. Thus $b \in Ha$ iff $ba^{-1} \in H$. Similarly $b \in aH$ iff $a^{-1}b \in H$.

Theorem 5.1.1 : Define a relation R_1 on G, denoted by $\sim$, by the rule $a \sim b$ if $b \in Ha$ (or equivalently by above lemma, $a \sim b$ if $ba^{-1} \in H$) where $a, b \in G$ and $H < G$. Then $\sim$ (R_1) is an equivalence relation on G.

Proof : (i) $a \sim a$ because $aa^{-1} = e \in H$, as $H < G$.   (Reflexive)

(ii) Now $a \sim b \Rightarrow ba^{-1} \in H$. But
$$ab^{-1} = (ba^{-1})^{-1} \text{ and } (ba^{-1})^{-1} \in H, \text{ as } ba^{-1} \in H$$
Thus $ab^{-1} \in H$ implying $b \sim a$.

So, $a \sim b \Rightarrow b \sim a$ (**Symmetric**)

(iii) Let $a, b, c \in G$ and $a \sim b, b \sim c$.

$a \sim b \Rightarrow ba^{-1} \in H$, $b \sim c \Rightarrow cb^{-1} \in H$

Consider ca^{-1}

$= cea^{-1} = c(b^{-1}b)a^{-1} = (cb^{-1})(ba^{-1}) \in H$

Thus $ca^{-1} \in H$ implying $a \sim c$

So, $a \sim b, b \sim c \Rightarrow a \sim c$ (Transitive)

Thus $\sim (R_1)$ is an equivalence relation on G.

By property of equivalent relations, $\sim (R_1)$ defined above will partition G into mutually disjoint equivalence classes. Let E_a be such an equivalence class.

Then,
$$E_a = \{b \in G : a \sim b\}$$
$$= \{b \in G : ba^{-1} \in H\}$$
$$= Ha \quad \text{(by lemma 5.1.1)},$$

showing that E_a is the right coset Ha.

Thus (i) the disjoint equivalence classes into which equivalence relation R_1 partitions G are the (distinct) right cosets $Ha, Hb, \ldots$ of H in G, and since any set is union of its equivalence classes, (ii) G is union of distinct right cosets of H in G.

Therefore the elements of G/H partition G or the set G/H is a partition of G.

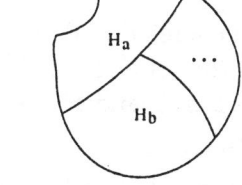

G partitioned by elements of G/H -

Corollary 5.1.1 : $Ha = Hb$ iff $ba^{-1} \in H$.

Proof : By equality of equivalence classes [Prelim. 2 (iv)], we know that

$$Ha = Hb$$
iff $\quad a \sim b$

which by definition of $a \sim b$ implies $b \in Ha$ or $ba^{-1} \in H$. Hence $Ha = Hb$ iff $ba^{-1} \in H$. Exactly proceeding by considering left cosets of H in G, we have analogous result in terms of left cosets.

Theorem 5.1.2 : Define a relation R_2 on G, denoted by $\sim$, by the rule $a \sim b$ if $b \in aH$, or equivalently (in view of lemma 5.1.1).

$a \sim b$ if $a^{-1}b \in H$, where $a, b \in G$ and $H < G$. Then $\sim$ (R_2) is an equivalence relation on G.

The proof is exactly similar to that of Theorem 5.1.2 and is left as an exercise for the readers.

The equivalence relation R_2 on G will partition G into mutually disjoint equivalence classes. As in the case of equivalence relation R_1, these equivalence classes are the (distinct) left cosets of H in G and hence G will be the union of distinct left cosets of H in G, or G will be the union of elements of the set G\H. Thus G\H is another partition of G.

G partitioned by elements of G\H.

Corollary 5.1.2 : $aH = bH$ iff $a^{-1}b \in H$

Corollary 5.1.3 : Choosing $a = e$ in corollaries 5.1.1 and 5.1.2, we get

(i) $Hb = He = H$ iff $b \in H$, and
(ii) $bH = eH = H$ iff $b \in H$

We have already come across these two results in illustrative examples 5.1.2.

We consider some more examples on sets G\H and G/H and their elements.

5.1.3 Illustrative Examples

1. For a fixed integer m, the set $m \cdot \mathbb{Z}$ is a subgroup of the group of integers $(\mathbb{Z}, +)$,

where $\qquad m \cdot \mathbb{Z} = \{\ldots\ -2m,\ -m,\ 0,\ m,\ 2m,\ \ldots\}$

Choose $m = 3$ to have a subgroup $H = 3\mathbb{Z}$

where $\qquad 3 \cdot \mathbb{Z} = \{\ldots\ -6,\ -3,\ 0,\ 3,\ 6,\ \ldots\}$

Then, left cosets of $3\mathbb{Z}$ in $\mathbb{Z}$ are:

$0\ (3\mathbb{Z}) = 0 + (3\mathbb{Z}) \quad$ as $* = +$ here

$\qquad = \{\ldots\ -6, -3, 0, 3, 6, \ldots\} = 3\mathbb{Z},$

$1\ (3\mathbb{Z}) = 1 + (3\mathbb{Z})$

$\qquad = \{\ldots\ 1+(-6), 1+(-3), 1+0, 1+3, 1+6, \ldots\}$

$\qquad = \{\ldots\ -5, -2, 1, 4, 7, \ldots\},$

$2\ (3\mathbb{Z}) = 2 + (3\mathbb{Z})$

$\qquad = \{\ldots\ 2+(-6), 2+(-3), 2+0, 2+3, 2+6, \ldots\}$

$\qquad = \{\ldots\ -4, -1, 2, 5, 8, \ldots\}$

$3\ (3\mathbb{Z}) = 3 + (3\mathbb{Z})$

$\qquad = \{\ldots\ 3+(-6), 3+(-3), 3+0, 3+3, 3+6, \ldots\}$

$\qquad = \{\ldots\ -3, 0, 3, 6, \ldots\}$

$\qquad = 3\mathbb{Z} \quad$ (This verifies corollary 5.1.3 (ii) that $b\mathrm{H} = \mathrm{H}$ iff $b \in \mathrm{H}$)

Similarly $4\ (3\mathbb{Z}) = \{\ldots\ -2, 1, 4, 7, \ldots\} = 1\ (3\mathbb{Z}), \ldots,$

$-1\ (3\mathbb{Z}) = \{\ldots\ -7, -4, -1, 2, \ldots\} = 2\ (3\mathbb{Z})$, etc.

Thus there are three distinct left cosets of $3\mathbb{Z}$ in $\mathbb{Z}$, namely $0\ (3\mathbb{Z})$ (which is $3\mathbb{Z}$ itself), $1\ (3\mathbb{Z})$ and $2\ (3\mathbb{Z})$ so that the left quotient set, $\mathbb{Z}\backslash 3\mathbb{Z} = \{0\ (3\mathbb{Z}), 1\ (3\mathbb{Z}), 2\ (3\mathbb{Z})\}$ and union of its elements

$$0\ (3\mathbb{Z}) \cup 1\ (3\mathbb{Z}) \cup 2\ (3\mathbb{Z})$$
$$= \mathbb{Z}$$

i.e. elements of $3\backslash 3\mathbb{Z}$ partition the group $\mathbb{Z}$.

The distinct right cosets of $3\mathbb{Z}$ in $\mathbb{Z}$ are

$$(3\mathbb{Z})\ 0 = 3\ \mathbb{Z},$$
$$(3\mathbb{Z})\ 1 = \{\ldots -5, -2, 1, 4, 7, \ldots\},\text{ and}$$
$$(3\mathbb{Z})\ 2 = \{\ldots -4, -1, 2, 5, \ldots\}.$$

Thus the right quotient set, $\mathbb{Z}/3\mathbb{Z} = \{(3\ \mathbb{Z})\ 0, (3\mathbb{Z})\ 1, (3\mathbb{Z})\ 2\}$ and union of its elements

$$(3\mathbb{Z})\ 0 \cup (3\mathbb{Z})\ 1 \cup (3\mathbb{Z})\ 2$$
$$= \mathbb{Z}$$

i.e. the set $\mathbb{Z}/3\mathbb{Z}$ is another partition of $\mathbb{Z}$.

Since $0\ (3\mathbb{Z}) = (3\mathbb{Z})\ 0$, $1\ (3\mathbb{Z}) = (3\mathbb{Z})\ 1$ and $2\ (3\mathbb{Z}) = (3\mathbb{Z})\ 2$, therefore the left quotient set $\mathbb{Z}\backslash 3\mathbb{Z}$ and the right quotient set $\mathbb{Z}/3\mathbb{Z}$ are identical, *i.e.* $\mathbb{Z}\backslash 3\mathbb{Z} = \mathbb{Z}/3\mathbb{Z}$ and their common value, denoted by $\mathbb{Z}\ |\ 3\mathbb{Z}$, is the quotient set of $\mathbb{Z}$ by $3\mathbb{Z}$. (Notice that $(\mathbb{Z}, +)$ is an abelian group).

In general for the subgroup $m\mathbb{Z}$ of $(\mathbb{Z}, +)$, where m is any positive integer,

$$\mathbb{Z}\backslash m\mathbb{Z} = \{0\ (m\mathbb{Z}), 1\ (m\mathbb{Z}), 2\ (m\mathbb{Z}), \ldots, (m-1)\ (m\mathbb{Z})\}$$
$$= \{r\ (m\mathbb{Z}) : r = 0, 1, \ldots (m-1)\}$$
$$= \mathbb{Z}/m\mathbb{Z}.$$

so that the quotient set $\mathbb{Z}\ |\ m\mathbb{Z}$ is given by $\mathbb{Z}\ |\ m = \{r\ (m\mathbb{Z}) : r = 0, 1, \ldots (m-1)\}$

Note : 1. Here the equivalence relation $\sim$ which partitiones $\mathbb{Z}$ into elements of quotient set $\mathbb{Z}\ |\ m\mathbb{Z}$ will be given by $a \sim b$ iff $ba^{-1} \in m\mathbb{Z}$. Since $a^{-1} = -a$ when $* = +$, $a \sim b$ iff $b - a \in m\mathbb{Z}$, which is same as $a \equiv b \pmod{m}$. Thus the elements of $\mathbb{Z}\ |\ m\mathbb{Z}$ are infact the residue classes of $\mathbb{Z}$ mod m.

2. As discussed in 5.1.2, example (1), $G\backslash H = \{1H, iH\}$ and $G/H = \{H1, Hi\}$, where $G = \{-1, 1, i, -i\}$ and $H = \{-1, 1\}$, $* = \cdot$ Also $G\backslash H = G/H$ so that $G\ |\ H = \{1H, iH\}$.

Similarly from 5.12, example (2),

Cosets and Lagrange's Theorem

$G\backslash H = \{\rho_0 H, \rho_1 H, \rho_2 H\}$ and $G/H = \{H\rho_0, H\rho_1, H\rho_2\}$ where $G = S_3 = \{\rho_0, \rho_1, \rho_2, \mu_1, \mu_2, \mu_3\}$ and $H = \{\rho_0, \mu_1\}$. As $\rho_1 H \neq H\rho_1$ and $\rho_2 H \neq H\rho_2$, $G\backslash H \neq G/H$ and the group S_3 has two partitions $G\backslash H$ and G/H.

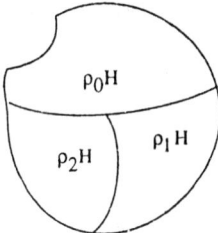
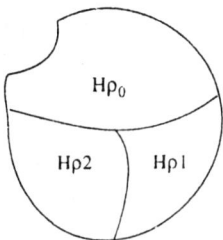

S_3 partitioned by elements of $S_3\backslash H$, $H = \{\rho_0, \mu_1\}$

S_3 partitioned by elements of S_3/H, $H = \{\rho_0, \mu_1\}$

3. Let $G = \mathbb{Z}_{12} = \{0, 1, 2, ..., 11\}$, $* = +_{12}$ and $H = <3> = \{0, 3, 6, 9\}$. Then $H < G$. Here $G\backslash H = \{oH, 1H, 2H\}$ and $G/H = \{H0, H1, H2\}$. Also $0H = Ho$, $1H = H1$ and $2H = H2$ (Note that group $\mathbb{Z}_{12}$ is abelian),

so that $\qquad G\backslash H = G/H = G \mid H$

and $\qquad G \mid H = \mathbb{Z}_{12} \mid <3>$

$\qquad\qquad\qquad = \{0H, 1H, 2H\}$, and

$\qquad OH \cup 1H \cup 2H = \mathbb{Z}_{12}$.

4. In case of the subgroup $A_3 = \{\rho_0, \rho_1, \rho_2\}$ of group $S_3 = \{\rho_0, \rho_1, \rho_2, \mu_1, \mu_2, \mu_3\}$, we will see that $S_3\backslash A_3 = S_3/A_3$ eventhough S_3 is a non-abelian group. $\rho_0 A_3$ and $\mu_1 A_3$ are two distinct left cosets of A_3 in S_3 where $\rho_0 A_3 = \{\rho_0, \rho_1, \rho_2\} = A_3$ and $\mu_1 A_3 = \{\mu_1 \rho_0, \mu_1 \rho_1, \mu_1 \rho_2\} = \{\mu_1, \mu_3, \mu_2\}$ so that $S_3\backslash A_3 = \{\rho_0 A_3, \mu_1 A_3\}$. The two distinct right cosets of A_3 in S_3 are $A_3 \rho_0$ and $A_3 \mu_1$, where $A_3 \rho_0 = \{\rho_0, \rho_1, \rho_2\} = A_3$ and $A_3 \mu_1 = \{\rho_0 \mu_1, \rho_1 \mu_1, \rho_2 \mu_1\} = \{\mu_1, \mu_2, \mu_3\} = \mu_1 A_3$ so that $S_3/A_3 = \{A_3 \rho_0, A_3 \mu_1\} = S_3\backslash A_3$.

Thus $S_3 \mid A_3 = \{\rho_0 A_3, \mu_1 A_3\}$.

5.2 General Properties of Cosets

We now state and prove some general properties possessed by cosets in the form of following theorems :

Theorem 5.2.1 : Let $H < G$ and $a, b \in G$. Then

(i) either $aH = bH$ or $aH \cap bH = \phi$,
(ii) either $H_a = H_b$ or $H_a \cap H_b = \phi$

In other words, any two left (or right) cosets of H in G are either identical or disjoint.

Proof: (i) We will prove that if $aH \cap bH \neq \phi$ then $aH = bH$. Let $c \in aH \cap bH$ which implies that $c \in aH$ and $c \in bH$.

$c \in aH \Rightarrow$ there is some h_1 in H s.t. $c = ah_1$ ...(1)
and $\quad c \in bH \Rightarrow$ there is some h_2 in H s.t. $c = bh_2$ ...(2)
so that, by (1) and (2),

$$ah_1 = bh_2$$
or, $\quad b = (ah_1)h_2^{-1}$ ...(3)
where $\quad h_1, h_2^{-1} \in H$

Now $bH \subseteq aH$, because for some $h \in H$, $bh \in bH$ and by (3),

$$bh = \{(ah_1)h_2^{-1}\}h$$
$$= a\{(h_1 h_2^{-1})h\} \in aH,$$

as $h_1, h_2^{-1}h \in H$. Similarly by taking out value of a in terms of b by (1) and (2), we can show that $aH \subseteq bH$. Thus aH and bH are subsets of each other, or $aH = bH$.

(ii) Proof is exactly on the lines of (i) above and is left as an exercise.

Theorem 5.2.2: Let $H < G$. Then:

(i) There exists an one-to-one correspondence between left coset aH and subgroup H $\forall$ $a \in G$.

(ii) There exists an one-to-one correspondence between right coset Ha and subgroup H $\forall$ $a \in G$, and

(iii) There exists an one-to-one correspondence between any two distinct left (or right) cosets aH and bH (or Ha and Hb).

Proof: (i) Define a map $f: H \to aH$ by the rule

$$hf = ah \ \forall \ h \in H \qquad ...(1)$$

Cosets and Lagrange's Theorem

Then the map (1) is well-defined (for $h_1, h_2 \in H$ if $h_1 = h_2$ then $h_1 f = ah_1 = ah_2 = h_2 f$ or $h_1 = h_2 \Rightarrow h_1 f = h_2 f$), one-to-one ($h_1 f = h_2 f \Rightarrow ah_1 = ah_2 \Rightarrow h_1 = h_2$ (By LCL) i.e. $h_1 f = h_2 f \Rightarrow h_1 = h_2$) and anto (for each $ah \in H$ there $h \in H$). Hence there is an one-to-one correspondence between aH and H.

(ii) Define the map $f : H \to Ha$ by $hf = ha$ and proceed exactly as in (i).

(iii) Define a map $f : aH \to bH$, where aH and bH are any two cosets of H in G and $a, b \in G$, by the rule

$$(ah)f = bh, \forall\ ah \in aH \qquad \ldots(2)$$

Then f is one-to-one because

$$(ah_1)f = (ah_2)f$$
$\Rightarrow \qquad bh_1 = bh_2$
$\Rightarrow \qquad h_1 = h_2$
$\Rightarrow \qquad ah_1 = ah_2$

and f is onto because for any element bh of bH, $h \in H$

$$\Rightarrow ah \in aH$$

and by definition of f, $(ah) f = bh$. (Prove yourself that map (2) is well-defined). This proves that map f defined by (2) is one-one onto and hence there is an one-one onto correspondence between any two left cosets aH and bH.

Similar is the proof when cosets are right cosets.

Note : By (i) and (ii) we conclude that 0 (aH) = 0 (H) and 0 (Ha) = 0 (H) and by (iii) that 0 (aH) = 0 (bH), and 0 (Ha) = 0 (Hb). Thus any two distinct left (or any two distinct right) cosets have the same number of elements if they are finite and the same cardinal number if they are infinite.

Theorem 5.2.3 : Let H < G. Then there is one-to-one correspondence between the sets G\H and G/H.

Proof : Define a map $f : G\backslash H \to G/H$ by the rule

$(aH) f = Ha^{-1}$ ...(1), $\forall\ aH \in G\backslash H$ where $a \in G$

(i) f is well-defined.

Let aH and bH be any two elements of $G\backslash H$ such that
$$aH = bH$$
$\Rightarrow \quad a^{-1}b \in H \quad$ (by cor. 5.1.2)
$\Rightarrow \quad Ha^{-1}b = H \quad (H < G$ so $Hh = H)$
$\Rightarrow \quad H(a^{-1}b)b^{-1} = Hb^{-1}$
$\Rightarrow \quad Ha^{-1}(bb^{-1}) = Hb^{-1}$
$\Rightarrow \quad Ha^{-1} = Hb^{-1}$
$\Rightarrow \quad (aH)f = (bH)f$

Thus, $aH = bH \Rightarrow (aH)f = (bH)f$ or f is well-defined.

(ii) f is one-to-one.

Consider $(aH)f = (bH)f$ where $(aH)f$ and $(bH)f \in G/H$. Then by (1),
$$Ha^{-1} = Hb^{-1}$$
$\Rightarrow \quad b^{-1}(a^{-1})^{-1} \in H \quad$ (by cor. 5.1.1)
$\Rightarrow \quad b^{-1}a \in H$
$\Rightarrow \quad b^{-1}aH = H \quad (H < G$ so $hH = H)$
$\Rightarrow \quad b(b^{-1}a)H = bH$
$\Rightarrow \quad (bb^{-1})aH = bH$
$\Rightarrow \quad aH = bH$

Thus $(aH)f = (bH)f$
$\Rightarrow \quad aH = bH$

or f is one-to-one.

(iii) f is onto.

For any right coset $Ha \in G/H$, there is a left coset $a^{-1}H \in G\backslash H$ such that
$$(a^{-1}H)f = H(a^{-1})^{-1} = Ha,$$
or f is onto.

Note: 1. By above theorem, the number of left cosets of H in G is same as the number of right cosets of H in G. If $G\backslash H$ and G/H are infinite sets then they have the same cardinality.

Cosets and Lagrange's Theorem

2. If G is an infinite group and H < G, it is possible that G\H (and hence G/H) are finite (see 5.1.3, example 1).

Theorem 5.2.4 : Let H < G. Then

(i) G is union of all left cosets of H in G.

(ii) G is union of all right cosets of H in G.

Proof : (i) By theorem 5.1.2 , the elements of set of all left cosets G\H partition G into mutually disjoint equivalence classes, hence by the property of equivalence classes, $G = \cup\, aH$, $\forall\, a \in G$.

Alternatively, if $x \in G$ then since $x = xe$ so that $x \in xH$ or any element x of G belongs to some left coset xH implying that $G = \cup\, aH$ $\forall\, a \in G$.

Similarly (ii) follows from Theorem 5.1.1 or you can construct alternative proof as above for (ii) also.

Theorem 5.2.4 takes us to the definition of *index* of a subgroup.

Definition 5.2.1 : The total number of distinct left (or right) cosets of H in G (H < G) is called the index of H in G, denoted by [G : H]. Thus

$$[G : H] = 0\,(G \backslash H) \text{ if set } G \backslash H \text{ is finite}$$

$$= 0\,(G/H) \quad \text{(by Th. 5.2.3)}$$

If G\H is infinite, index [G : H] is defined as the cardinal number of G\H or the cardinal number of G/H.

Note : 1. Every subgroup of a finite group will have finite index.

2. Index of a subgroup may be finite even if group is infinite.

Examples : From 5.1.3, $[\mathbb{Z} : 3\mathbb{Z}] = 3$, $[\mathbb{Z} : m\mathbb{Z}] = m$; for $G = \{-1, 1, i, -i\}$ and its $H = \{-1, 1\}$, $[G : H] = 2$;

for $G = S_3$ and its $H = [\rho_0, \mu_1]$, $[G : H] = 3$, for $G = S_3$ and its subgroup $H = \{\rho_0, \rho_1, \rho_2\} = A_3$, $[G : H] = [S_3 : A_3] = 2$ and for $G = \mathbb{Z}_{12}$ and its $H = <3>$, $[G : H] = 3$.

5.3 Lagrange's Theorem for Finite Groups and Its Applications

By making use of properties discussed in 5.2, we will now prove a powerful theorem due to Lagrange* on order of a finite group which is one of the oldest one in theory of groups and contributes significantly in the study of finite groups. It has very important consequences as well.

5.3.1 Lagrange's Theorem

The order of subgroup of a finite group divides the order of group. Thus if $0(G)$ is finite and $H < G$ then $0(H)$ divides $0(G)$.

Proof: Let $0(G) = n$, $H = \{h_1, h_2, ..., h_m\}$ be a subgroup of G so that $0(H) = m$. (obviously $m \leq n$)

To prove that: m divides n

Take an arbitrary element $a_1 \in G$. Then $a_1 H$ is a left coset of H in G and $0(a_1 H) = 0(H) = m$ (Thm. 5.2.2 (*i*))

---

* *Joseph Louis Lagrange (1736-1813)* : Born on January 25, 1736 in Turin (Italy), the great French analyst Lagrange is one of the great mathematicians of all time. He worked in diverse areas of mathematics such as theory of equations, differential equations, number theory, celestial mechanics and fluid mechanics. Such was his outstanding ability that at the age of 16, he became Professor of Mathematics in the Royal School of Artillery at Turin. A great admirer of Maclaurin (1698-1746), Lagrange presented his famous theorem on order of finite groups in 1770. His memoir "Réflections sur la théorie algé brique des équations", which he submitted to the Berlin Academy contains a general method of solving polynomials of degree greater than 4, proved to be a valuable source for the development of group theory.

Cosets and Lagrange's Theorem 199

Let there be k distinct left cosets of H in G and they be $a_1H, a_2H, ..., a_kH$. Then for any i and j ($1 \leq i < j \leq k$), $a_iH \cap a_jH = \phi$ (Thm 5.2.1 (i)),

$$0(a_iH) = 0(a_jH) = m \text{ and}$$
$$G = a_1H \cup a_2H \cup ... \cup a_kH \quad \text{(Thm 5.2.4 (i))}$$

$\Rightarrow \quad 0(G) = 0\{a_1H \cup a_2H \cup ... a_kH\}$
$$= 0(a_1H) + 0(a_2H) + ... + 0(a_kH)$$
$$= m + m + ... + m \quad (k \text{ times})$$

$\Rightarrow \quad n = km$

$\Rightarrow \quad \dfrac{n}{m} = k,\quad$ where k, the number of distinct left cosets of H in G, is a positive integer.

$\Rightarrow$ m divides n, or $0(H)$ divides $0(G)$.

Note : 1. In the above proof, which is due to Camille Jordan (1838-1922), right cosets of H in G can be used in place of left casets.

2. $\dfrac{n}{m} = k \Rightarrow$ Number of distinct left (or right) cosets of H in G = $\dfrac{0(G)}{0(H)}$.

5.3.2 Deductions from Lagrange's Theorem

Corollary 5.3.1 : Order of an element of a finite group divides the order of group.

Proof : Let $0(G) = n$ and a be an arbitrary element of G. Construct a cyclic subgroup H of G by collecting all integral powers of the element a so that $H = <a>$.

Then $\qquad 0(H) = 0(a) \qquad\qquad\qquad$ ...(1)

Since $H < G$, by Lagrange's theorem, $0(H)$ divides $0(G)$ or by (1) $0(a)$ divides $0(G)$.

Corollary 5.3.2 : A group whose order is a prime number is cyclic and hence abelian group.

Proof : Let $0(G) = p$, where p is prime, and a be an arbitrary element of G. Construct a cyclic subgroup H of G by collecting all integral powers of a so that H = <a>. Since H < G, by Lagranges theorem, $0(H)$ divides $0(G)$ which is a prime number p. Thus $0(H)$ is either 1 or p. $0(H) \neq 1$ as a is an arbitrary element so that $0(H) = p$ or H = G. Then G = <a> is a cyclic group. Since all cyclic groups are abelian, G is an abelian group. (Also H = <a> $\Rightarrow$ $0(H) = 0(a)$, *i.e.* order of every non-identity element of a group of prime order p is also p.)

Corollary 5.3.3 : A group whose order is a prime number has no proper subgroup.

Proof : By Lagrange's theorem, order of any subgroup of a group will divide order of group. Thus if G is a group whose order is a prime number p then order of any subgroup of G can have orders 1 or p, *i.e.* subgroups of G will be improper subgroups and G can not have any proper subgroup.

Corollary 5.3.4 : If G is a finite group of order n, then $\forall\, a \in G$, $a^n = e$.

Proof : Let $0(a) = m$ so that

$$m \leq n, \text{ and } a^m = e \qquad ...(1)$$

By corollary 5.3.1, $0(a)$ divides $0(G)$, *i.e.* m divides n. Let t be a positive integer and

$$\frac{0(G)}{0(a)} \text{ or } \frac{n}{m} = t \text{ or } n = mt \qquad ...(2)$$

By (1) and (2)

$$a^n = a^{mt}$$
$$= (a^m)^t = (e)^t = e$$

Corollary 5.3.5 : If G is a finite group and H < G, then index of H in G, $[G : H] = \dfrac{0(G)}{0(H)}$.

Cosets and Lagrange's Theorem

Proof : In the proof of Lagrange's theorem, by note (2), number of distinct left (or right) cosets of H in G = $\frac{0(G)}{0(H)}$, which in conjunction with the definition of index of H in G implies [G : H] = $\frac{0(G)}{0(H)}$.

5.3.3 Converse of Lagrange's Theorem

If G is a group of order n and m is a divisor of n then a natural question which arises here is that : Does G always have a subgroup of order m ? The answer, in general, is No. Thus converse of Lagrange's theorem is not necessarily true, *i.e.* even if m divides the order n of a group G, it does not necessarily imply that G would have a subgroup of order m. The following examples illustrate the situation and tell us that, for abelian groups, the converse of Lagrange's theorem is true while for non-abelian groups it is not necessarily true.

Examples an Converse of Lagrange's Theorem : 1. Consider the (non-abelian) alternating group A_4 where $0(A_4) = 12$. Although 6 is a divisor of 12, A_4 has no subgroup of order 6. (See also example on 5.3.4). Thus converse of Lagrange's theorem does not hold true in the case of group A_4. (See the group table of A_4 in 2.3.2)

2. Consider the (non-abelian) octic group D_4 where $0(D_4) = 8$, and $D_4 = \{\rho_0, \nu_1, \nu_2, \nu_3, \tau_1, \tau_2, \tau_3, \tau_4\}$. 2 is a divisor of 8 and D_4 has a subgroup $\{\rho_0, \tau_1\}$ of order 2. 4 is a divisor of 8 and D_4 has a subgroup $\{\rho_0, \nu_1, \nu_2, \nu_3\}$ of order 4. (See lattice diagram for D_4 for all of its subgroups). Thus the converse of Lagrange's theorem holds true in the case of group D_4.

3. Consider the abelian group $\mathbb{Z}_6 = \{0, 1, 2, 3, 4, 5\}$ where $* = +_6$, and $0(\mathbb{Z}_6) = 6$. 2 is a divisor of 6 and $\mathbb{Z}_6$ has a subgroup $\{0, 3\}$ of order 2. 3 is a divisor of 6 and $\mathbb{Z}_6$ has a subgroup $\{0, 2, 4\}$ of order 3. Thus the converse of Lagrange's theorem holds true in the case of abelian group $\mathbb{Z}_6$.

4. The converse of Lagrange's theorem also holds good in case of non-abelian group S_3 (order 6) and (abelian) Klein's group K_4 (order 4).

Note : Lagrange's theorem is also helpful in finding nature of subgroups of a finite group. For example, consider the group S_3. Since $0\ (S_3) = 6$, by Lagrange's theorem its possible subgroups will have 2 or 3 elements, both of which are prime numbers. Hence by corollary 5.3.2, all the possible subgroups of S_3 will be cyclic.

5.3.4 Sylow p-Subgroups

In general the converse of Lagrange's theorem is not true. However we have Sylow's theorem which is closely related with the converse of Lagrange's theorem. Here we state (without proof) the theorem :

Sylow's theorem : If p is a prime number and p^m is the highest power of p which divides n, the order of G, then G has atleast one subgroup of order p^m.

Any such subgroup of G is called a *Sylow p-subgroup* of G, after the name of L. Sylow[*] who proved the theorem in 1873.

Example : According to Sylow's theorem, the alternating group A_4 with $n = 12$ should have atleast one subgroup each of orders 2^2 (*i.e.* 4) and 3^1 (*i.e.* 3). From chapter 3, problem 3.11 we have seen that A_4 has a subgroup $V = \{\rho_0, \tau_1, \tau_3, v_2\}$ of order 4 and subgroups $H_1 = \{\rho_0, -l, m_1\}$, $H_2 = \{\rho_0, l, m_4\}$, $H_3 = \{\rho_0, l_2, m_2\}$ and $H_4 = \{\rho_0, l_3, m_3\}$, each of order 3. Thus V is Sylow 2-subgroup and H_1, H_2, H_3 and H_4 are Sylow 3-subgroups of the group A_4.

---

[*] *Peter Ludwig Sylow (1832-1918)* : Well-known for his contributions on structural results in finite groups, Sylow was born in Norway on December 12, 1832. His famous research paper published in Maths Ann. 5 (1872) pp. 584-94 consists of eight theorems which are now known as Sylow's theorems. The paper also contains result on number of Sylow p-subgroups possessed by a finite group and the result showing that any two Sylow p-subgroups are conjugate. His results are useful in classifying groups of a given order.

5.3.5 Applications of Lagrange's Theorem

We have seen in previous sections that Lagrange's theorem provides informations about groups of prime as well as non-prime orders. Corollary 5.3.4 helps us in proving a useful theorem, Fermat's theorem (also called Fermat's little theorem) which has found applications in number theory namely in cryptography (a branch associated with writings in secret codes) and primality testing.

Theorem 5.3.1 (Euler's theorem) : Let m be a positive integer and p be an integer relatively prime to m, then $p^{\phi(m)} \equiv 1$ (mod m), where $\phi(m)$ is the Euler's ϕ-function.

Proof : Consider the group $G = \{\bar{p} : p \in \mathbb{Z} \text{ and } gcd(p, m) = 1\}$

where $\bar{p}$ denotes the residue class of $\mathbb{Z}$ (mod m) and $* =$ multiplication of residue classes. Since $\bar{p} \in G$, by Cor. 5.3.4,

$$\bar{p}^{0(G)} = \bar{1}, \text{ where the residue class } \bar{1} \text{ is the identity } e \text{ of } G$$

$\Rightarrow \quad \bar{p}^{\phi(m)} = \bar{1} \quad (0(G) = \phi(m))$

$\Rightarrow \quad \bar{p} \cdot \bar{p} \cdot \bar{p} \ldots \phi(m) \text{ times} = \bar{1}$

$\Rightarrow \quad \overline{p \cdot p \cdot p \ldots \phi(m) \text{ times}} = \bar{1} \quad ((\bar{r}\bar{s} = \overline{rs})$

$\Rightarrow \quad \overline{p^{\phi(m)}} = \bar{1}$

$\Rightarrow \quad p^{\phi(m)} \equiv 1 \pmod{m}.$

Theorem 5.3.2 (Fermat's theorem) : If n is any integer and p is a prime number then $n^p \equiv n \pmod{p}$.

Proof : Since p is a prime number, $\phi(p) = p - 1$, and $gcd(n, p) = 1$.

By Euler's theorem

$$n^{\phi(p)} \equiv 1 \pmod{p}$$

$\Rightarrow \quad n^{p-1} \equiv 1 \pmod{p}$

$\Rightarrow \quad n^p \equiv n \pmod{p}$

Fermat's theorem can also be proved without using Euler's theorem.

Fermat's theorem (alternative proof): (i) If n is divisible by p then $n^p - n$ is also divisible by p, i.e. if $n \equiv 0 \pmod{p}$ then $n^p - n \equiv 0 \pmod{p} \Rightarrow n^p \equiv n \pmod{p}$ which proves the theorem in this case.

(ii) If n is not divisible by p, i.e. if $n \not\equiv 0 \pmod{p}$ then $\bar{n} \neq \bar{0}$ and $\bar{n} \in G$, where $G = \{\bar{1}, \bar{2}, ..., \overline{p-1}\}$ is the group of non-zero residue classes modulo the prime p and $* =$ multiplication of residue classes, and $0(G) = p - 1$

By cor. 5.3.4,

$$\bar{n}^{0(G)} = \bar{1} \quad (\text{Here } e = \bar{1})$$

$\Rightarrow \qquad \bar{n}^{(p-1)} = \bar{1}$

$\Rightarrow \qquad n^{p-1} \equiv 1 \pmod{p}$

$\Rightarrow \qquad n^p \equiv n \pmod{p}$

Conserve of Fermat's theorem is not true, *i.e.* $n^l \equiv n \pmod{l}$ $\not\Rightarrow l$ is a prime number. One such example is

$$n^{5.61} \equiv n \pmod{561} \qquad ...(1)$$

for any integer n. Non-prime numbers like 561 satisfying (1) are called *Carmichael numbers*. Give example of some more Carmichael numbers.

Application to number theory: The theorems of Fermat and Euler are very useful theorems. Fermat's theorem can be used to obtain remainders when n^p is divided by p. We give here an example of obtaining remainder using a particular form of Fermat's theorem.

Example: Since $197 \equiv 2 \pmod 5$, we have

$\overline{197} = \bar{2}$ (in the group $(\mathbb{Z}_4, X_5)$ where $\mathbb{Z}_4 = \{1, 2, 3, 4\}$)...(1)

Also, in $(\mathbb{Z}_4, X_5)$
$$0(\bar{2}) = 4,$$
so $\bar{2}^{33} = \bar{2}^1$ (Theorem 4.2.2 (i) chapter 4)
...(2)

By (1) and (2),
$$\overline{197}^{33} = \bar{2}^{33} = \bar{2}^1,$$
i.e. when 197^{33} is divided by 5, remainder is 2.

Solved Problems Set on Chapter 5

Problem 5.1 : For the subgroup $H = \{\rho_0, l, m_1\}$ of the group A_4, find $A_4 \backslash H$ and A_4/H.

Solution : From the group table of A_4, the distinct left cosets of H in A_4 are $\rho_0 H$, $\tau_1 H$, $v_2 H$ and $\tau_3 H$ where
$$\tau_1 H = \{\tau_1 * \rho_0, \tau_1 * l, \tau_1 * m_1\} = \{\tau_1, l_3, m_4\},$$
etc. and thus
$$A_4 \backslash H = \{\rho_0 H, \tau_1 H, v_2 H, \tau_3 H\}$$

The distinct right cosets of H in A_4 are $H\rho_0$, $H\tau_1$, Hv_2 and $H\tau_3$, where
$$H\tau_1 = \{\rho_0 * \tau_1, l * \tau_1, m_1 * \tau_1\} = \{\tau_1, l_1, m_3\},$$
etc. and thus
$$A_4/H = \{H\rho_0, H\tau_1, Hv_2, H\tau_3\}.$$
Here $A_4 \backslash H \neq A_4/H$

Problem 5.2 : For the subgroup $H = \{\rho_0, v_1, v_2, v_3\}$ of the group D_4, find $D_4 \backslash H$ and D_4/H.

Solution : There are two distinct left cosets of H in D_4, namely $\rho_0 H$ and $\tau_1 H$, where
$$\tau_1 H = \{\tau_1 * \rho_0, \tau_1 * v_1, \tau_1 * v_2, \tau_1 * v_3\} = \{\tau_1, \tau_2, \tau_3, \tau_4\}$$
Thus $D_4 \backslash H = \{\rho_0 H, \tau_1 H\}$

Similarly work out D_4/H. Is $D_4 \backslash H = D_4/H$?

Problem 5.3 : Show that a non-abelian group has atleast six elements.

Solution : (i) The one element group $\{e\}$ where $*$ is given by $e * e = e$ is an abelian group.

(*ii*) The groups having 2, 3 and 5 elements are all abelian groups as they are prime ordered groups. (Corollary 5.3.2).

(*iii*) Let $0(G) = 4$ and $e \neq a \in G$. Then, by Lagrange's theorem, the subgroup $<a>$ of G can have order 2 or 4. If $0 <a> = 4$ then $<a> = G$ and hence G is also cyclic and therefore abelian. Thus groups having 4 elements are abelian.

(*iv*) Finally $0(S_3) = 6$ and S_3 is a non-abelian group.

Thus from (*i*) to (*iv*), a non-abelian group should have atleast six elements.

Problem 5.4 : In a group G, define $F = \{a \in G : aH = Ha,$ where $H < G\}$. Prove that $F < G$.

Solution : See problem 3.4.

Problem 5.5 : Let G be a finite group and H and K be two subgroups of G.

(a) If $0(H) = i$ and $0(K) = j$ where $gcd(i, j) = 1$, prove that $H \cap K = \{e\}$

(b) If $0(H) = 0(K) = p$ where p is a prime number and H, K are distinct, prove that $H \cap K = \{e\}$.

Solution : (a) Since $H < G$, $K < G$ thus $H \cap K < G$. Further, $H \cap K \subset H \Rightarrow H \cap K < H$, and $H \cap K \subset K \Rightarrow H \cap K < K$. Thus by Lagrange's theorem $0(H \cap K)$ will divide i, the $0(H)$ and $0(H \cap K)$ will also divide j, the $0(K)$, so that $0(H \cap K)$ will be a common factor of i and j. But, given that $gcd(i, j) = 1$, therefore

$$0(H \cap K) = 1 \quad \text{or} \quad H \cap K = \{e\}.$$

(b) Since $H \cap K < H$ $(H \cap K \subset H)$, by Lagranges theorem $0(H \cap K)$ will divide p, the $0(H)$. But given that p is a prime number, thus $0(H \cap K) = 1$ or p.

If $0(H \cap K) = p$, $H \cap K = H$ which implies that $H \subseteq K$, which is not possible as, given that $0(H) = 0(K)$ and H and K are distinct.

Thus $\quad 0(H \cap K) = 1 \quad \text{or} \quad H \cap K = \{e\}$

Problem 5.6: Let G be a group and $a, b \in G$ such that $0(a) = i$, and $0(b) = j$ where $\gcd(i, j) = 1$. If $ab = ba$, prove that $0(ab) = ij$.

Solution: In problem 5.5, take $H = \langle a \rangle$ and $K = \langle b \rangle$ so that we have

$$\langle a \rangle \cap \langle b \rangle = \{e\}$$

By solved problem 1.36 (a),

$$0(ab) = \text{LCM}(i, j)$$
$$= ij \quad (\gcd(i, j) = 1)$$

Problem 5.7: Let H and K be any two finite subgroups of a group G then prove that

$$0(HK) = \frac{0(H)\, 0(K)}{0(H \cap K)}$$

In particular, if $H \cap K = \{e\}$, then $0(HK) = 0(H)\, 0(K)$.

Proof: Let $I = H \cap K$ so that $I < G$ and $I \subseteq H$ and $I \subseteq K$.

Also $I < H$ where H is finite,

thus index of I in H, $[H : I]$, is finite, say j.

There will be j distinct left cosets of I in H, say $a_1 I, a_2 I, \ldots, a_j I$ ($a_1, a_2, \ldots, a_j \in H$)

and $\qquad H = a_1 I \cup a_2 I \cup \ldots \cup a_j I$

Then $\qquad HK = (a_1 I \cup a_2 I \cup \ldots \cup a_j I) K$

$\qquad\qquad\quad = a_1 K \cup a_2 K \cup \ldots \cup a_j K^* \quad (I \subseteq K)$

$\Rightarrow \quad 0(HK) = 0(a_1 K) + 0(a_2 K) + \ldots + 0(a_j K)$

$\qquad\qquad\quad = 0(K) + 0(K) + \ldots + 0(K)\ j$ times

$\qquad\qquad\qquad\qquad\qquad (0(a_1 K) = 0(K))$

$\qquad\qquad\quad = j \cdot 0(K)$

$\Rightarrow \quad 0(HK) = [H : I]\, 0(K)$

$$= \frac{0(H)}{0(I)} \cdot 0(K)$$

$$= \frac{0(H) \, 0(K)}{0(H \cap K)}$$

In particular, if $H \cap K = \{e\}$ then $0(H \cap K) = 1$ and we have

$$0(HK) = 0(H) \, 0(K)$$

(* The cosets $a_1 K, a_2 K, \ldots, a_j K$ are distinct left cosets of K otherwise $a_1 I, a_2 I, \ldots, a_j I$ where $I = H \cap K$ will not be distinct).

Review Exercise on Chapter 5

1. For the group A_4 and its subgroup $V = \{\rho_0, v_2, \tau_1, \tau_3\}$, find $A_4 \backslash V$ and A_4 / V. Are they equal?

2. Let $G = D_4$ and $H = \langle \tau_4 \rangle = \{\rho_0, \tau_4\}$. Find $D_4 \backslash H$ and D_4 / H.

3. For the improper subgroups $H_1 = \{\rho_0\}$ and $H_2 = A_3$ of the group A_3, compute left and right quotient sets $A_3 \backslash H_1, A_3 \backslash H_2, A_3 / H_1$ and A_3 / H_2.

4. If $aH = bH$, prove that $b^{-1}a \in H$.

    [*Hint*: $aH = bH \Rightarrow a^{-1}b \in H \Rightarrow (a^{-1}b)^{-1} \in H$ $(H < G)$]

5. Show by means of examples that the converse of Lagranges theorem holds true in the finite groups S_3, S_4 and D_n $(n \geq 3)$.

6. If p is a prime number, show that every group of order p^2 is abelian.

7. If H is a Sylow p-subgroup of a finite group G and $a \in G$, prove that aHa^{-1} is also a Sylow p-subgroup of G.

8. Let H_1 and H_2 be subgroups of finite indices of a group G. Prove that $H_1 \cap H_2$ is also a subgroup of G of finite order. (Poincare's theorem)

    (*Hint*: Show that, $a(H_1 \cap H_2) = (aH_1) \cap (aH_2)$)

9. Let G be a finite group and H, K be two subgroups of G such that $K \subset H$. Verify by an example that

    $[G : K] = [G : H] [H : K]$

6. NORMAL SUBGROUPS AND QUOTIENT GROUPS

6.1 Normal Subgroups

If $(G, *)$ is an abelian group and H is a subgroup of G then by commutative law, $a * H = H * a$, for all $a \in G$ ...(1)

i.e. the left coset determined by an element of an abelian group coincides with its right coset. It happens that for some subgroups, (1) holds even if G is non-abelian. Subgroups H of a group G having characteristic (1) are called *Normal subgroups* of G. We denote a normal subgroup H of a group G by $H \triangleleft G$. The study of normal subgroups helps in determining the group structure as well as the nature of homomorphisms having domain G. In terms of left and right quotient sets, we define a normal subgroup H as :

Definition 6.1.1 : A subgroup H of a group G is normal iff $G\backslash H = G/H$, where $G\backslash H$ and G/H are respectively the sets of left and right cosets of H in G. The following theorem establishes the equivalence of this definition with (1).

Theorem 6.1.1 : The subgroup H of a group G is normal iff $aH = Ha$, for all $a \in G$.

Proof : If $H \triangleleft G$, by definition 6.1.1,

$$aH = Hb \quad ...(1), \quad \text{where } a, b \in G.$$

Since $a = ae$, thus $a \in aH$ and by (1),

$$a \in Hb \Leftrightarrow ab^{-1} \in H \quad \text{(Note 4, chapter 5)}$$

$\Leftrightarrow Hab^{-1} = H$ (Cor. 5.1.3)

$\Leftrightarrow Ha = Hb$

$= aH$ (by 1)

Thus $H \triangleleft G \Leftrightarrow Ha = aH$, for any $a \in G$.

Theorem 6.1.1 yields another *characterization* for normal subgroups as follows :

Theorem 6.1.2 : A subgroup H or a group G is normal iff

(i) $aha^{-1} \in H$, or equivalently

(ii) $a^{-1}ha \in H$, where $a \in G$ and $h \in H$.

Proof : Since $H \triangleleft G$,

$\Leftrightarrow \quad\quad aH = Ha$, for any $a \in G$

$\Leftrightarrow \quad\quad aHa^{-1} = H$, which by equality of sets implies

$\quad\quad\quad aha^{-1} \in H$, where $h \in H$.

Similarly, by pre-operating by a^{-1} in the defining relation $aH = Ha$, we get $H = a^{-1}Ha$ which further implies $a^{-1}ha \in H$, for $h \in H$.

Corollary 6.1.1 : $H \triangleleft G$ iff $aHa^{-1} = H$ or equivalently $a^{-1}Ha = H$.

Note : 1. If H is subgroup of an abelian group G, $aH = Ha$, for all $a \in G$ and thus $G\backslash H = G/H$, so that subgroups of an abelian group are all normal (subgroups).

2. When $H \triangleleft G$, we do not differentiate between a left coset aH and a right coset Ha; we simply call aH (or Ha) a coset.

3. We will apply condition (i) (or condition (ii)) of Th. 6.1.2 to test whether a subgroup H is normal or not.

4. From solved problem 1.2, chapter 3, set $H^a = a^{-1}Ha$, where $a \in G$, forms a subgroup of G, called conjugate subgroup to H in G. Thus if $H \triangleleft G$, by cor. 6.1.1, subgroup H coincides with all of its conjugate subgroups in G. Thus a normal subgroup is also called a *self conjugate subgroup*.

Normal Subgroups and Quotient Groups

5. Since $aea^{-1} = e$ for all $a \in G$, $\{e\} \triangleleft G$ and for any $a, x \in G$, since $axa^{-1} \in G$, $G \triangleleft G$. Thus the singleton $\{e\}$ and entire group G are normal subgroups of a group G.

Definition 6.1.2 : The improper subgroups $\{e\}$ and G of a group G are always normal and they are called *Improper* (or *trivial*) normal subgroups. Remaining all normal subgroups of G, if exist, are called *proper* (or *non-trivial*) normal subgroups.

Definition 6.1.3 : A group G which has no normal subgroups except the two improper normal subgroups $\{e\}$ and G is called a *Simple group*.

Examples : 1. Finite simple groups are significant in the study of all finite groups. A complete list of finite simple groups, which are infinite in number, have recently been arranged and all the alternating groups A_n for $n \geq 5$ are shown to be simple. (The proof that A_5, A_6, ... are simple is well beyond the scope of this book.)

2. By Lagrange's theorem, a group of prime order has no proper subgroup and therefore has no proper normal subgroup. Thus groups of prime orders are all simple groups.

We give below yet another characteristic of normal subgroups in terms of index (See, definition 5.2.1), which helps us in testing a subgroup for its normalty.

Theorem 6.1.3 : A subgroup H of a group G with index 2 is a normal subgroup of G.

Proof : Since $[G : H] = 2$, there are two left cosets and two right cosets of H in G. For $a \in G$, let the two left cosets be eH and aH and the two right cosets be He and Ha. Let $a \notin H$ so that $eH \neq aH$ and $He \neq Ha$. We know that,

$$G = eH \cup aH \qquad \ldots(1),$$
and
$$G = He \cup Ha \qquad \ldots(2)$$
so that
$$eH \cup aH = He \cup Ha$$
$$\Rightarrow \qquad aH = Ha, \text{ for any } a \in G$$
$$\Rightarrow \qquad H \triangleleft G.$$

Definition 6.1.4 : Let N be a proper normal subgroup of a group G. Then N is said to be *maximal normal subgroup* iff there does not exist any other normal subgroup M of G such that N $\subset$ M.

6.1.1 Illustrative Examples

1 Since ($\mathbb{Z}$, +) and Klein's groups are abelian, all of their subgroups are normal. Thus for $n \in \mathbb{Z}$, $n\mathbb{Z} \triangleleft \mathbb{Z}$ and for Klein's group K_4, $\{e, a\} \triangleleft K_4$, $\{e, b\} \triangleleft K_4$, $\{e, c\} \triangleleft K_4$.

2. The testing for normality can be done for the subgroups $A_3 = \{\rho_0, \rho_1, \rho_2\}$ and $H_1 = \{\rho_0, \mu_1\}$ of (non-abelian) group S_3 as follows :

Consider $\mu_2^{-1}\rho_1\mu_2$,

where
$$\mu_2 = \begin{pmatrix} 1 & 2 & 3 \\ 3 & 2 & 1 \end{pmatrix} \in S_3,$$

and
$$\rho_1 = \begin{pmatrix} 1 & 2 & 3 \\ 2 & 3 & 1 \end{pmatrix} \in A_3$$

$$= \begin{pmatrix} 3 & 2 & 1 \\ 1 & 2 & 3 \end{pmatrix} \begin{pmatrix} 1 & 2 & 3 \\ 2 & 3 & 1 \end{pmatrix} \begin{pmatrix} 1 & 2 & 3 \\ 3 & 2 & 1 \end{pmatrix}$$

$$= \begin{pmatrix} 1 & 2 & 3 \\ 3 & 1 & 2 \end{pmatrix} = \rho_2 \in A_3$$

Thus for $\mu_2 \in S_3$ and $\rho_1 \in A_3$, $\mu_2^{-1}\rho_1\mu_2 \in A_3$, by Th. 6.1.2 (ii) $A_3 \triangleleft S_3$.

Again, for $\rho_1 \in S_3$ and $\mu_1 \in H_1$,

Consider $\rho_1^{-1}\mu_1\rho_1$

$$= \begin{pmatrix} 2 & 3 & 1 \\ 1 & 2 & 3 \end{pmatrix} \begin{pmatrix} 1 & 2 & 3 \\ 1 & 3 & 2 \end{pmatrix} \begin{pmatrix} 1 & 2 & 3 \\ 2 & 3 & 1 \end{pmatrix} = \begin{pmatrix} 1 & 2 & 3 \\ 3 & 2 & 1 \end{pmatrix} = \mu_2 \notin H_1$$

so that $H_1 \not\triangleleft S_3$.

Normal Subgroups and Quotient Groups 213

Similarly the subgroups $H_2 = \{\rho_0, \mu_2\}$ and $H_3 = \{\rho_0, \mu_3\}$ are not normal subgroups of S_3.

From problem 3.3, we know that $A_n < S_n$.

3. The alternating group A_n is normal subgroup of the group S_3, as

$$[S_n : A_n] = \frac{0(S_n)}{0(A_n)} = \frac{\lfloor n}{\frac{1}{2}(\lfloor n)},$$

hence by Th. 6.1.3, $A_n \triangleleft S_n$. In fact $A_n \triangleleft S_n$ for all $n \in \{2, 3, 4, \ldots\}$ due to the same reason.

4. We will apply cor. 6.1.1 to test if the subgroup $V = \{\rho_0, v_2, \tau_1, \tau_3\}$ of S_4 is normal.

Let $a = (1\ 2\ 3) \in S_4$. Then
$aVa^{-1} = \{a\rho_0 a^{-1}, av_2 a^{-1}, a\tau_1 a^{-1}, a\tau_3 a^{-1}\}$

By actual calculations,

$a\rho_0 a^{-1} = aa^{-1} = \rho_0$

$$av_2 a^{-1} = \begin{pmatrix} 1 & 2 & 3 & 4 \\ 2 & 3 & 1 & 4 \end{pmatrix} \begin{pmatrix} 1 & 2 & 3 & 4 \\ 3 & 4 & 1 & 2 \end{pmatrix} \begin{pmatrix} 2 & 3 & 1 & 4 \\ 1 & 2 & 3 & 4 \end{pmatrix}$$

$$= \begin{pmatrix} 1 & 2 & 3 & 4 \\ 4 & 3 & 2 & 1 \end{pmatrix} = \tau_1,$$

Similarly

$$a\tau_1 a^{-1} = \begin{pmatrix} 1 & 2 & 3 & 4 \\ 2 & 3 & 1 & 4 \end{pmatrix} \begin{pmatrix} 1 & 2 & 3 & 4 \\ 4 & 3 & 2 & 1 \end{pmatrix} \begin{pmatrix} 2 & 3 & 1 & 4 \\ 1 & 2 & 3 & 4 \end{pmatrix}$$

$$= \begin{pmatrix} 1 & 2 & 3 & 4 \\ 2 & 1 & 4 & 3 \end{pmatrix} = \tau_3$$

and

$$a\tau_3 a^{-1} = \begin{pmatrix} 1 & 2 & 3 & 4 \\ 2 & 3 & 1 & 4 \end{pmatrix} \begin{pmatrix} 1 & 2 & 3 & 4 \\ 2 & 1 & 4 & 3 \end{pmatrix} \begin{pmatrix} 2 & 3 & 1 & 4 \\ 1 & 2 & 3 & 4 \end{pmatrix}$$

$$= \begin{pmatrix} 1 & 2 & 3 & 4 \\ 3 & 4 & 1 & 2 \end{pmatrix} = v_2$$

Thus, $aVa^{-1} = \{\rho_0, \tau_1, \tau_3, v_2\} = V$, i.e.
by cor. 6.1.1 $V \triangleleft S_4$.

Similarly it can be shown that $V \triangleleft A_4$. Thus we have the series

$$\{\rho_0\} \subset V \subset A_4 \subset S_4,$$

in which each is normal subgroup of the one which follows it.

5. Consider the Hamiltonian group (see, solved problem 1.13) which is non-abelian and has order 8. By Lagranges theorem, its possible subgroups will be of orders 4 and 2. For a subgroup H of order 4, since

$$[G : H] \neq \frac{0(G)}{0(H)} = 2,$$ hence subgroups of order 4 will be normal.

The subgroup of order 2 is $K = \{-1, 1\}$ for which you can easily show that

$$aKa^{-1} = K, \quad \text{for any } a \in G$$

Thus $K \triangleleft G$.

Thus all the subgroups of this non-abelian group are normal.

6. Consider the group $GL_n(\mathbb{R})$ and its proper subgroup $SL_n(\mathbb{R})$. Take $A \in GL_n(\mathbb{R})$ and $X \in SL_n(\mathbb{R})$ so that A and X are $n \times n$ non-singular matrices and det X, $|X| = 1$. Since $|A \times A^{-1}| = 1$ and $A \times A^{-1} = X$ therefore $SL_n(\mathbb{R}) \triangleleft GL_n(\mathbb{R})$.

6.2 Quotient (or Factor) Group

Let $(G, *)$ be any group.

If $N \triangleleft G$, we have seen that $G \backslash N = G/N$. Let $G \backslash N = G/N = G \mid N$, the quotient set which consists of all distinct cosets of N in G (see below definition 5.1.3) Thus,

$G \mid N = \{aN : a \in G\}$ (or, $G \mid N = \{Na : a \in G\}$)

Normal Subgroups and Quotient Groups

Notice that aN stands for $a * N$, (similarly Na for $N * a$). Our objective now is to construct a group of all distinct cosets of N in G for which we require a b.o. over the set G | N. The following theorem provides the same.

Theorem 6.2.1 : If G is a group and N a subgroup of G then N is normal in G iff product of any two left cosets of N in G is again a left coset of N in G.

Proof : Given $N < G$.

(i) *First part* : Assume that $N \triangleleft G$.

Then, for any $a, b \in G$, aN and bN are two left cosets of N in G and $(aN) \cdot (bN)$ is their product.

Now $(a * N) \cdot (b * N)$

$= (aN)(bN) = a(Nb)N = a(bN)N = ab(NN) = ab\,N$

(as $N \triangleleft G$, $Nb = bN$ and $NN = N$)

Since $ab \in G$, thus abN is a left coset of N in G. Thus product of two left cosets is a left coset.

(ii) *Second part* : Assume that the product of two left cosets aN and bN of N in G is a left coset, where $a, b \in G$. Set $b = a^{-1}$, we see that $(aN)(a^{-1}N)$ is a left coset of N in G.

Since $\quad (ae)(a^{-1}e) \in (aN)(a^{-1}N) \quad$ (as $N < G$)

$\Rightarrow aa^{-1} \quad$ or $\quad e \in (aN)(a^{-1}N)$

Also, $e \in eN$, where eN is a left coset of N in G. Thus the cosets $(aN)(a^{-1}N)$ and eN have e as a common element. Thus by theorem 5.2.1

$(aN)(a^{-1}N) = N, \quad a \in G$

$\Rightarrow \quad (an_1)(a^{-1}n_2) \in N, \quad$ where $n_1, n_2 \in N$

$\Rightarrow \quad \{(an_1)(a^{-1}n_2)\}\,n_2^{-1} \in N \quad$ (as $n_2^{-1} \in N$, $N < G$)

$\Rightarrow \quad an_1a^{-1} \in N, \quad$ where $a \in G$ and $n_1 \in N$

$\Rightarrow \quad N \triangleleft G$.

Thus if the product $(aN)(bN)$ is a left coset then $N \triangleleft G$.

Note : 1. Above theorem holds true for right cosets as well.

2. If the operation in group G is + (addition), the "product" $(aN)(bN)$ will be

$(a + N) + (b + N) = (a + b) + N.$

3. From the first part of above theorem, the product of two left cosets is defined by

$$(aN)(bN) = ab\ N$$

where $N \triangleleft G$ and abN is a left coset of N in G, which is in fact $(a*N) \cdot (b*N) = (a*b)N$, where $*$ is b.o. in G. Thus if $aN, bN \in G \mid N$, then

$$abN \in G \mid N,$$

where $N \triangleleft G$. Hence product of cosets, as defined above, is a b.o. over $G \mid N$, where $N \triangleleft G$.

Theorem 6.2.2 : If $N \triangleleft G$, the quotient set $G \mid N$ together with the operation of product of cosets satisfies all group conditions.

Proof : Let for $a, b, c, \ldots \in G$, $aN, bN, cN, \ldots$ be distinct cosets (as $N \triangleleft G$) of N in G so that $G \mid N = \{aN, bN, cN, \ldots\}$. The operation over $G \mid N$ is given to be the product of cosets defined by

$$(aN)(bN) = abN, \quad \text{for any } a, b \in G$$

By Th. 6.2.1, $abN \in G \mid N$ hence the operation is a b.o. over $G \mid N$. Group axioms for $G \mid N$–

(i) Associative law : For any $a, b, c \in G$

$$(aN)[(bN)(cN)] = (aN)[bcN] = a(bc)N$$
$$= (ab)cN$$
$$= (ab)N\,(cN)$$
$$= [(aN)(bN)](cN)$$

(ii) Existence of id. element in $G \mid N$:

Since $eN \in G \mid N$ and for any $aN \in G \mid N$, $(eN)(aN) = (ea)N = aN$

$\Rightarrow eN$ is the identity element of the set $G \mid N$.

(iii) Existence of inverses in $G \mid N$:

For any $a \in G$, $a^{-1} \in G$ so that

$$aN, a^{-1}N \in G \mid N$$

Since $(a^{-1}N)(aN) = (a^{-1}a)N = eN$, the id. element of $G \mid N$.

Normal Subgroups and Quotient Groups

Thus $(a^{-1}N)$ is the inverse of aN, *i.e.* each element of $G \mid N$ has its inverse in $G \mid N$.

Thus the set of all cosets $G \mid N$, where $N \triangleleft G$, forms a group under product of cosets as a b.o.

Definition 6.2.1 : The group $G \mid N$ is called *factor* or *quotient group*.

Definition 6.2.2 : Order of factor group $G \mid N$ is equal to the number of distinct cosets of N in G.

If G is a finite group, by above definition and Lagrange's theorem

$$0 \ (G \mid N) = \frac{0 \ (G)}{0 \ (N)} = [G : N]$$

6.2.1 Illustrative Examples

1 (*i*) The group $G = \{1, -1, i, -i\}$ is an abelian group under b.o. of ordinary multiplication. Hence its subgroup $H = \{-1, 1\}$ is normal subgroup of G so that $G \mid H$ is a factor group. From (1) of I.E. 5.1.2, we have $G \mid H = \{1H, iH\}$. Notice that $G \mid H$ has two elements namely the cosets $1H$ and iH so that $0 \ (G \mid H) = 2$ as confirmed by the formula,

$$0 \ (G \mid H) = \frac{0 \ (G)}{0 \ (H)} = \frac{4}{2} = 2.$$

(*ii*) $H = <3>$ is a subgroup of abelian group $(\mathbb{Z}_{12}, +_{12})$ and hence $H \triangleleft \mathbb{Z}_{12}$. Thus $\mathbb{Z}_{12} \mid <3>$ is a factor group. From (3) of I.E. 5.1.3, $\mathbb{Z}_{12} \mid <3> = \{0H, 1H, 2H\}$ so that

$$0 \ (\mathbb{Z}_{12} \mid <3>) = 3 = \frac{0 \ (\mathbb{Z}_{12})}{0 \ <3>}.$$

(*iii*) $A_3 \triangleleft S_3$ and from (4) of I.E. 5.1.3, factor group $S_3 \mid A_3 \ \{\rho_0 A_3, \mu_1 A_3\}$,

so that $\qquad 0 \ (S_3 \mid A_3) = 2 = \dfrac{0 \ (S_3)}{0 \ (A_3)}.$

2. Since $(\mathbb{Z}, +)$ is an infinite abelian group and $3\mathbb{Z} < \mathbb{Z}$ so that $3\mathbb{Z} \triangleleft \mathbb{Z}$ and $\mathbb{Z} \mid 3\mathbb{Z}$ is a factor group which contains all

distinct cosets of $3\mathbb{Z}$ in $\mathbb{Z}$. From (1) of I.E. 5.1.3, there are three distinct cosets of $3\mathbb{Z}$ in $\mathbb{Z}$, namely $0\ (3\mathbb{Z})$, $1\ (3\mathbb{Z})$ and $2\ (3\mathbb{Z})$, so that the factor group,

$$\mathbb{Z}\mid 3\mathbb{Z} = \{0\ (3\mathbb{Z}),\ 1\ (3\mathbb{Z}),\ 2\ (3\mathbb{Z})\}$$

and $\quad 0\ (\mathbb{Z}\mid 3\mathbb{Z}) = 3$

(The formula $0\ (G\mid N) = \dfrac{0\ (G)}{0\ (N)}$ is not applicable here as $\mathbb{Z}$ is an infinite group.)

In general, as in (1) of I.E. 5.1.3, when $m \in \mathbb{Z}^+$,

$$\mathbb{Z}\mid m\mathbb{Z} = \{r\ (m\mathbb{Z}) : r = 0, 1, 2, \ldots (m-1)\}$$

so that $\quad 0\ (\mathbb{Z}\mid m\mathbb{Z}) = m$.

3. From (4) of I.E. 6.1.1 $V = \{\rho_0, v_2, \tau_1, \tau_3\}$ is a normal subgroup of A_4, so that $A_4 \mid V$ is a factor group. Similarly $V \triangleleft S_4$, $S_4 \mid V$ is also a factor group.

There are three distinct cosets of V in A_4 namely $V\rho_0$, Vl and Vm_1, so that $A_4 \mid V = \{V\rho_0, Vl, Vm_1\}$ and $0\ (A_4 \mid V) = 3$, as confirmed by $0\ (A_4 \mid V) = \dfrac{0\ (A_4)}{0\ (V)} = \dfrac{12}{4}$.

Let $Vl = C$ and $Vm_1 = D$. Also $V\rho_0 = V$, we have following group table for the factor group A_4/V :

*	V	C	D
V	V	C	D
C	C	D	V
D	D	V	C

(* is coset multiplication)

Observe that :

(i) Index $[A_4 : V] = 3$ and $V \triangleleft A_4$.

(ii) Factor group A_4/V is abelian, its order being 3 but the group A_4 is non-abelian; exactly as in the case of factor group $S_3 \mid A_3$.

Normal Subgroups and Quotient Groups 219

Theorem 6.2.3 : Factor group of an abelian group is abelian.

Proof : Leg G be an abelian group and $N \triangleleft G$. Take two elements, aN and bN of $G \mid N$.

Then, by definition of product of two cosets

$(aN)(bN) = abN$

$\qquad = baN$ (as $a, b \in G$ and G is abelian)

$\qquad = (bN)(aN)$,

i.e. $G \mid N$ is abelian.

Thus if group G is abelian, factor group $G \mid N$ is also abelian.

Converse of theorem 6.2.3 is not true, because as in (3) I.E. 6.2.1, factor group $A_4 \mid V$ is abelian but not the group A_4. Similarly the factor group $S_3 \mid A_3$ is abelian but not the group S_3 (See 1 (*iii*) of I.E. 6.2.1). Thus if factor group $G \mid N$ is abelian, it does not imply that G is also abelian.

Solved Problems Set

Problem 6.1 : Prove that the intersection of two normal subgroups of a group G is a normal subgroup of G.

Solution : Let H and K be two normal subgroups of G. $H \cap K < G$ as $H < G$ and $K < G$. To prove that $H \cap K$ is normal, let $a \in G$ and $n \in H \cap K$ so that $n \in H$ and $n \in K$. Now $a \in G$, $n \in H$ and $H \triangleleft G$,

Therefore $\qquad ana^{-1} \in H \qquad$ ...(1)

Similarly $\qquad a \in G, n \in K$ and $K \triangleleft G$,

$\qquad\qquad ana^{-1} \in K \qquad$ ...(2)

By (1) and (2),

$\qquad ana^{-1} \in H \cap K$,

where $a \in G$ and $n \in H \cap K$.

Thus $H \cap K \triangleleft G$.

(The proof can be extended for intersection of any number of normal subgroups).

Problem 6.2 : If N and H are two subgroups of a group G such that $N \triangleleft G$ and $N \subset H$, then prove that $N \triangleleft H$. Give an example of this situation.

Solution : Let $h \in H$. Since $H < G$, $h \in G$. Now $N \triangleleft G$,
Therefore $\quad h^{-1}Nh = N$
Thus $\quad h \in H \Rightarrow h^{-1}Nh = N \quad \ldots(1)$
and since h is arbitrary element of H, by (1) $N \triangleleft H$.

Example : $V = \{\rho_0, \tau_1, \tau_3, v_2\}$ and A_4 are two subgroups of S_4 where $V \triangleleft S_4$ (See (4) of I.E. 6.1.1) and $V \subset A_4$. Then by above problem $V \triangleleft A_4$, as confirmed by (4), I.E. 6.1.1.

Problem 6.3 : Show that centre Z of a group G is a normal subgroup of G.

Solution : By definition,
$$Z = \{g \in G : gx = xg, \text{ for all } x \in G\}$$

We know that $Z < G$ (see theorem 3.4.2). To show that Z is normal as well, consider the composition of elements xgx^{-1} where $x \in G$ and, $g \in Z$. Now
$$xgx^{-1} = (xg) x^{-1} = (gx) x^{-1} = g \in Z$$

Thus for $x \in G$ and $g \in Z$, $xgx^{-1} \in Z$ which implies that $Z \triangleleft G$.

Example : The centre of Hamiltonian group is $\{-1, 1\}$ which is shown to be normal subgroup in (5) of I.E. 6.1.1.

Problem 6.4 : If $N \triangleleft G$ and $H < G$ then prove the following :

(i) $\quad H \cap N \triangleleft H$

(ii) $\quad NH = HN$

(iii) $\quad NH < G$, and

(iv) $\quad N \triangleleft HN$

Solution : Given that G is a group, $N \triangleleft G$ and $H < G$.

(i) $H \cap N < G$ as $H < G$ and $N < G$, and also $H \cap N \subseteq H$. Thus we have $H \cap N \subseteq H$ where $H < G$, which implies that $H \cap N < H$ as $H \cap N < G$.

Normal Subgroups and Quotient Groups

Consider the elements $h \in H$ and $l \in H \cap N$ then $l \in H$ and $l \in N$.

Now $N \triangleleft G$, we have normality condition

$$hlh^{-1} \in N \qquad \ldots(1)$$

Also $\qquad h \in H, l \in H$ and $H < G$

$\Rightarrow \qquad hl \in H$. Now $h^{-1} \in H$

$\Rightarrow \qquad (hl)h^{-1} \in H \qquad \ldots(2)$

By (1) and (2), $hlh^{-1} \in H \cap N$

Thus we showed that $h \in H$, $l \in H \cap N$ them

$$hlh^{-1} \in H \cap N \Rightarrow H \cap N \triangleleft H.$$

(ii) We will show that $NH \subset HN$ as well as $HN \subset NH$.

Let $nh \in NH$, where $n \in N, h \in H$

But we can write

$$nh = e\,(nh), \quad \text{where} \quad e \text{ is id. element of } G$$
$$= (hh^{-1})(nh) \quad (e = hh^{-1})$$
$$= h\,(h^{-1}nh)$$

Since $h^{-1}nh \in N$ as $N \triangleleft G$, $nh = hn_1$ where $n_1 = h^{-1}nh \in N$, so that $hn_1 \in HN$. Thus $nh \in HN$, i.e. $nh \in NH \Rightarrow nh \in HN$, or $NH \subseteq HN$.

Similarly the other part that $HN \subseteq NH$.

Thus combining, we get $NH = HN$, where $N \triangleleft G$ and $H < G$. ($NH = HN$ means normal subgroups of a group commute with the complexes of the group.)

(iii) We know that $H < G$, $K < G$ then $HK < G$ iff $HK = KH$.

Here $N \triangleleft G$ so that $N < G$ and $H < G$ and by part (ii) above $NH = HN$, thus $NH < G$.

(iv) Since any element $n \in N$ can be written as $n = en$ and since $en \in HN$, as $e \in H$ ($H < G$), thus $n \in N \Rightarrow n \in HN$ i.e. $N \subseteq HN$.

By part *(iii)* HN < G; and also N < G such that N ⊆ HN, thus N < HN. For normalty of N, consider the elements $h_1 n_1$ ∈ HN and n ∈ N, so that h_1 ∈ H and n_1 ∈ N. Since N ⊲ G and n_1, n ∈ G, we have

$$n_1 n n_1 \in N \qquad \ldots(1)$$

Now, for $h_1 n_1$ ∈ HN and n ∈ N,

$$(h_1 n_1)\, n\, (h_1 n_1)^{-1} = (h_1 n_1) n (n_1^{-1} h_1^{-1})$$
$$= h_1 (n_1 n n_1^{-1}) h_1^{-1} \in N \text{ as } N \triangleleft G$$

and h_1 ∈ G, $n_1 n n_1^{-1}$ ∈ N, by (1)

Thus for any $h_1 n_1$ ∈ HN and n ∈ N, $(h_1 n_1)\, n\, (h_1 n_1)^{-1}$ ∈ N

⇒ N ⊲ HN

Problem 6.5 : If N ⊲ G, M ⊲ G then show that the product set NM ⊲ G.

Solution : M ⊲ G ⇒ M < G

By prob. 6.4 *(ii)*, a normal subgroup commutes with a complex, thus NM = MN so that NM < G.

For normality of NM, consider the elements g ∈ G and nm ∈ NM, where n ∈ N and m ∈ M. Now $g\,(nm)\,g^{-1}$

$$= g\,(nem)g^{-1} = g\,(n\,(g^{-1}g)m)g^{-1}$$
$$= (gng^{-1})\,(gmg^{-1}) \qquad \ldots(1)$$

Since N ⊲ G, g ∈ G and n ∈ N ⇒ gng^{-1} ∈ N

and M ⊲ G, g ∈ G and m ∈ M ⇒ gmg^{-1} ∈ M,

we have $(gng^{-1})(gmg^{-1})$ ∈ NM

so that by (1), $g\,(nm)\,g^{-1}$ ∈ NM, where g ∈ G and nm ∈ NM. thus NM ⊲ G.

Example : Let N = {e, a} and M = {e, b}. Then N ⊲ K_4 and M ⊲ K_4, where K_4 is Klein's group. Now

$$NM = \{e * e, e * b, a * e, a * b\} = \{e, b, a, c\} = K_4$$

where K_4 is (improper) normal subgroup of K_4. Thus $NM \triangleleft K_4$.

Problem 6.6: If $N \triangleleft G$, $M \triangleleft G$ and that $N \cap M = \{e\}$ then prove that for all $n \in N$ and $m \in M$, $nm = mn$.

Solution: Consider the composition of elements $nmn^{-1}m^{-1}$ where $n \in N$ and $m \in M$ and hence $n^{-1} \in N$, $m^{-1} \in M$ as $N, M < G$.

Since $N \triangleleft G$ and $m \in G$, $n^{-1} \in N$, by normality condition of N,

$$mn^{-1}m^{-1} \in N$$

$\Rightarrow \quad n(mn^{-1}m^{-1}) \in N \quad \ldots(1)$

Again, $M \triangleleft G$ and $n \in G$, $m \in M$, by normality condition of M,

$$nmn^{-1} \in M$$

$\Rightarrow \quad (nmn^{-1})m^{-1} \in M \quad \ldots(2)$

By (1) and (2),

$$nmn^{-1}m^{-1} \in N \cap M$$

But given that $N \cap M = \{e\}$,

thus $\quad nmn^{-1}m^{-1} = e$

$\Rightarrow \quad nmn^{-1} = m$

$\Rightarrow \quad nm = mn$, where $n \in N$ and $m \in M$.

Thus when N and M are two normal subgroups of a group G having only e in their intersection then every element of N commutes with each element of M. The above example from K_4 illustrates this problem. If, in addition, G = NM then G is called the *internal direct product* of N and M.

Problem 6.7: Let G be a group and H be only subgroup of G of finite order, then prove that $H \triangleleft G$.

Solution: Let $0(H) = m$ where H is the only subgroup of G of order m. If $x \in G$, we know that

$$xHx^{-1} < G$$

We will first show that $0(xHx^{-1}) = 0(H)$

Let $H = \{h_1, h_2, ..., h_m\}$

Then, $xHx^{-1} = \{xh_1x^{-1}, xh_2x^{-1}, ..., xh_mx^{-1}\}$

No two elements in the set xHx^{-1} are same, for if for any

$$xh_ix^{-1} = xh_jx^{-1}, \text{ where } 1 \leq i \leq m \; 1 \leq j \leq m$$

$\Rightarrow$ $h_i = h_j$ (by cancellation laws),

which is not possible as $0(H) = m$

Thus $0(xHx^{-1}) = m$

$\Rightarrow \quad xHx^{-1} = H$

$\Rightarrow \quad xH = Hx, \text{ for any } x \in G$

$\Rightarrow \quad H \triangleleft G$

Problem 6.8 : Show that factor group of a finite cyclic group is also cyclic.

Solution : Let G be a finite cyclic group of order m and

$$G = \{a^m = e, a^1, a^2, ..., a^{m-1}\}, \cdot G = <a>$$

Since cyclic groups are abelian, all subgroups of G are normal. Let $N \triangleleft G$, so that G | N is a factor group. Let $l \in G \mid N$. Then $l = bN$, where $b \in G$. Since $G = <a>$, $b = a^k$ where $1 \leq k < m$.

Thus $l = (a^k) N$

$= (a * a * ... k \text{ times}) N$

$= (a * N) * (a * N) ... k \text{ times}$

$= (aN)(aN) ... k \text{ times}$

$= (aN)^k$

i.e. any element l of G | N is expressed as integral power of aN. Thus G | N is cyclic and G | N = $<aN>$.

Note : 1. Since $a^kN = (aN)^k$ for any integer k, above result is true for any cyclic group - finite or infinite.

2. *Converse* of this result is not true, since the factor group

$$S_3 \mid A_3 = \{\rho_0 A_3, \mu_1 A_3\}$$
$$= \langle \mu_1 A_3 \rangle$$

is cyclic but not the group S_3.

Problem 6.9 : If $a \in G$ and $N \triangleleft G$ then prove that $0\,(Na)$ is divisor of $0\,(a)$, where $Na \in G \mid N$.

Solution : Let $\quad 0\,(a) = n \Rightarrow a^n = e$

$\Rightarrow \qquad\qquad Na^n = Ne = N \quad ...(1)$

Since $\qquad\qquad Na^n = (Na)^n,\;$ (as shown in prob. 6.8 that
$\qquad\qquad\qquad\qquad\qquad a^k N = (aN)^k$)

$\Rightarrow \qquad\qquad N = (Na)^n \quad$ (by 1)

But N is the id. element of $G/N \Rightarrow 0\,(Na) \leq n$ and $0\,(Na)$ will be a divisor of n.

Problem 6.10 : If $N \triangleleft G$, $H \triangleleft G$ such that $N \subset H$, then prove that $H \mid N \triangleleft G \mid N$.

Solution : Since $N \triangleleft G$, $H \triangleleft G$ and $N \subset H$, therefore $N \triangleleft H$ and $H \mid N$ is factor group.

Let $aN \in H \mid N$. Then $a \in H \Rightarrow a \in G \Rightarrow aN \in G \mid N$

Then $\qquad aN \in H \mid N \Rightarrow aN \in G \mid N \cdot$

$\Rightarrow \qquad\qquad H \mid N \subseteq G \mid N$

$\Rightarrow \qquad\qquad H \mid N < G \mid N \quad$ (as $H \mid N$ is a group)

For normality of $H \mid N$ in $G \mid N$, consider the composition of elements $(gN)\,(hN)\,(gN)^{-1}$, where $gN \in G \mid N$ and hence $g \in G$, and $hN \in H \mid N$ and hence $h \in H$.

Now $(gN)\,(hN)\,(gN)^{-1}$

$\qquad\qquad\qquad = \{(gN)\,(hN)\}\,(N^{-1}g^{-1})$

$\qquad\qquad\qquad = (ghN)\,(N^{-1}g^{-1}) \quad$ (as $N \triangleleft G$, product
$\qquad\qquad\qquad\qquad\qquad\qquad\qquad (gN)\,(hN) = gh\,N$)

$\qquad\qquad\qquad = (ghN)\,(Ng^{-1})$

$\qquad\qquad\qquad = (ghN)\,(g^{-1}N) \quad$ (as $N \triangleleft G$, $N^{-1} = N$
$\qquad\qquad\qquad\qquad\qquad\qquad\qquad$ and $Ng^{-1} = g^{-1}N$)

$\qquad\qquad\qquad = ghg^{-1}N \qquad\qquad\qquad ...(1)$

Let $ghg^{-1} = h_1$, then $h_1 \in H$ as $H \triangleleft G$ and hence $ghg^{-1} \in H$ for some $g \in G$ and $h \in H$.

Thus by (1)
$$(gN)(hN)(gN)^{-1} = h_1 N, \text{ where } h_1 \in H$$
Since $h_1 N \in H \mid N$
$\Rightarrow \quad (gN)(hN)(gN)^{-1} \in H \mid N$, for some
$\quad gN \in G \mid N$ and $hN \in H \mid N$,
$\Rightarrow \quad H \mid N \triangleleft G \mid N$

Problem 6.11 : If G is a group with centre Z and $a \in Z$ then show that $<a> \triangleleft G$.

Solution : By definition
$$Z = \{g \in G : gx = xg, \text{ for all } x \in G\}$$

Let $H = <a>$, the cyclic subgroup of G generated by element a. If $h \in H$, then for some integer n, $h = a^n$. Consider the composition of elements yhy^{-1}, where $y \in G$ and $h \in H$, to test the normalty of H in G.

Now $\quad yhy^{-1}$
$\quad\quad = ya^n y^{-1} = (yay^{-1})^n$, for any integer n
$\quad\quad = (ayy^{-1})^n$ (given that $a \in Z$, so $ay = ya$ for any $y \in G$)
$\quad\quad = a^n \in H$

Thus for any $y \in G$ and $h \in H$, we showed that $yhy^{-1} \in H$
$\Rightarrow H \triangleleft G$ or $<a> \triangleleft G$.

Problem 6.12 : If G is a group and N(a) is the normaliser of its element a, then show that $<a> \triangleleft N(a)$.

Solution : By definition
$$N(a) = \{x \in G : ax = xa\}$$

Let $H = <a>$, the cyclic subgroup of G generated by element a. If $h \in H$, then for some $n \in \mathbb{Z}$, $h = a^n$. Also $h \in H \Rightarrow h \in G$ and $ah = aa^n = a^{n+1} = a^n a = ha$
$\Rightarrow \quad\quad\quad\quad h \in N(a)$.

Thus $h \in H \Rightarrow h \in N(a)$ so that $H \subseteq N(a)$. We know that $N(a) < G$. Also $H < G$ such that $H \subseteq N(a)$, hence $H < N(a)$. Consider the composition of elements yhy^{-1}, where $y \in N(a)$ and $h \in H$, for normality of H in $N(a)$. Now

$$yhy^{-1} = ya^ny^{-1} = (yay^{-1})^n, \text{ for any integer } n.$$
$$= (ayy^{-1})^n, \text{ as } y \in N(a)$$
$$= a^n \in H$$

Thus for any $y \in N(a)$ and $h \in H$, we showed that $yhy^{-1} \in H$ $\Rightarrow H \triangleleft N(a)$.

Review Exercise on Chapter 6

1. Show that $H = \{\rho_0, \tau_3\}$ is a normal subgroup of $V = \{\rho_0, v_2, \tau_1, \tau_3\}$ where V is a normal subgroup of A_4, but H is not a normal subgroup of A_4.
2. If $H < G$, $K < G$ then show that $aH \cap aK = a(H \cap K)$.
3. If $H \triangleleft G$ then show that for any $a \in G$ and for any integer n, $(Ha)^n = Ha^n$.
4. If G is an abelian group and $H = \{a \in G : 0(a) \text{ is finite}\}$, then prove that $H \triangleleft G$.
5. If G is a finite group and n is a positive integer such that for any $a, b \in G$, $(ab)^n = a^nb^n$ then prove that the sets

   $N = \{a \in G : a^n = e\}$ and

   $M = \{a^n : a \in G\}$ are normal subgroups of G.
6. If G is a finite group and $H < G$, $K < G$ then prove that $[G : H]$ is a factor of $[G : K]$.
7. If $H < K$ and $K < G$ such that G/K and K/H are finite sets then show that

   $0(G/H) = 0(G/K) \cdot 0(K/H)$.
8. If $H < G$ such that $[G : H] = 2$, then prove that $G \mid H$ is isomorphic to a cyclic group of order 2.
9. Find order and all the elements of the factor group $(\mathbb{R}, +) \mid (\mathbb{Z}, +)$.
10. Give an example to illustrate the problem 6.10.

7. MAPPINGS IN GROUP THEORY

In 1.9 of chapter 1, while discussing elementary properties of a group, we have defined homomorphism (*hmp*) and isomorphism (*ismp*) between two groups. In the present chapter we study them in detail.

7.1 Types of homomorphism

In chapter 1, onto-*hmp* and one-one onto *hmp* are, respectively, named as Epimorphism and Isomorphism. Other special types of homomorphism are defined as follows :

Definition 7.1.1 : A homomorphism $\sigma : G \to G'$ between two groups G and G' is said to be a *monomorphism* if σ is one-to-one as well.

Thus an isomorphism between two groups G and G' is a monomorphism as well as an epimorphism.

Definition 7.1.2 : A homomorphism $\sigma : G \to G$ is said to be an *endomorphism*. Thus an endomorphism is a *hmp* of a group G to itself.

Definition 7.1.3 : An isomorphism $\sigma : G \to G$ is said to be an *automorphism*.

Thus an automorphism is an isomorphism of a group G to itself.

Definition 7.1.4 : The special automorphism $i_a : G \to G$ defined by $b i_a = a b a^{-1}$, for all $b \in G$, and $a \in G$ is called an *inner automorphism* determined by the element a.

Definition 7.1.5 : (Kernel of a homomorphism) Let $\sigma : G \to$ G' be a homomorphism and e and e' be, respectively, the

Mappings in Group Theory

identity elements of G and G'. Kernel of σ, denoted by K_σ, is the set defined as : $K_\sigma = \{x \in G : x\sigma = e^1\}$.

Thus Kernel K_σ is the set of all those elements of group G which are carried over to the id. element e' of group G' by the *hmp* σ. Obviously $K_\sigma \subseteq G$. We will also see that K_σ is never an empty set.

7.1.1 Illustrative Examples

(1) The map $\sigma : (\mathbb{Z}, +) \to (\{\pm 1, \pm i\}, \cdot)$ defined by $m\sigma = i^m$ for all $m \in \mathbb{Z}$, is an epimorphism. Here $e = 0$ and $e' = 1$, and since $(4m)\,\sigma = 1$, $m \in \mathbb{Z}$, therefore Kernel $K_\sigma = 4\,\mathbb{Z} = \{..., -8, -4, 0, 4, 8, ...\}$.

(2) The map $\sigma : (\mathbb{C}, +) \to (\mathbb{R}, +)$ defined by $(a + ib)\,\sigma = a$, for all $a + ib \in \mathbb{C}$, is an epimorphism and $K_\sigma = \{ib : b \in \mathbb{R}\}$ i.e., the Kernel consists of all those complex numbers whose real parts are zero.

(3) The map $\sigma_1 : (Q^*, \cdot) \to (Q^+, \cdot)$ defined by $p\sigma_1 = |p|$ for all $p \in Q^*$ is an onto-*hmp* (epimorphism) and $K\sigma_1 = \{-1, 1\}$. The map $\sigma_2 : (\mathbb{C}^*, \cdot) \to (\mathbb{R}^*, \cdot)$ defined by $z\,\sigma_2 = |z|$ for all $z \in \mathbb{C}^*$ is an into-*hmp* whose Kernel consists of all complex numbers with their modulus equal to 1.

(4) The map $\sigma_1 : (\mathbb{Z}, +) \to (\{\pm 1\}, \cdot)$ defined by

$$m\sigma_1 = \begin{cases} 1 \text{ when } m \text{ is even or zero} \\ 0 \text{ when } m \text{ is odd} \end{cases}$$

for all $m \in \mathbb{Z}$. Then σ_1 is an epimorphism but not a monomorphism. Since $e' = 1$, $K\sigma_1$ is the set of all even integers including zero.

The map $\sigma_2 : (\mathbb{R}, +) \to (\{e^{i\theta} : \theta \in \mathbb{R}\}, \cdot)$ defined by $x\,\sigma_2 = e^{ix}$ for all $x \in \mathbb{R}$ is also an epimorphism but not a monomorphism. What is $K\sigma_2$?

(5) The map $\sigma : (\mathbb{Z}, +) \to (\mathbb{Z}, +)$ defined by $m\sigma = mn$, for all $m \in \mathbb{Z}$, where n is a fixed integer, is an endomorphism which is one-to-one (monomorphism) but not an onto (Epimorphism) if $n \neq \pm 1$.

(6) The map $\sigma : G \to G$, where G is any group and σ is defined by $x\sigma = e$ for all $x \in G$, is an endomorphism. Here $K_\sigma = G$.

(7) The map $\sigma : (\mathbb{C}^*, \cdot) \to (\mathbb{C}^*, \cdot)$ defined by $z\sigma = z^n$, for all $z \in \mathbb{C}^*$, where $n \in \mathbb{Z}^+$, is an endomorphism. Since
$$(e^{2r\pi i/n})\sigma = (e^{2r\pi i/n})^n, \text{ by definition of } \sigma,$$
$$= e^{2r\pi i}$$
$$= \cos 2r\pi + i \sin 2r\pi$$
$$= 1, \text{ the id. element of } \mathbb{C}^*, \text{ and further,}$$
$z\sigma = 1 \Rightarrow z^n = 1 \Rightarrow z = 1^{1/n} \Rightarrow z$ is one of nth roots of unity, thus $K_\sigma = \{e^{2r\pi i/n} : r = 0, 1, \ldots n-1\}$ i.e., K_σ is set of n n^{th} roots of unity.

(8) In 2.3.2 (ii) of chapter-2, we have noticed that the Dihedral group D_3 of symmetries of an equilateral triangle and the group S_3 of all permutations of first three natural numbers 1, 2, 3 are the same. This 'sameness' comes from the following isomorphism which exists between D_3 and S_3 :

Define the map $\sigma : D_3 \to S_3$ as

$\text{rot}_0 \to \rho_0$, $\text{rot}_{\frac{2\pi}{3}} \to \rho_1$, $\text{rot}_{\frac{4\pi}{3}} \to \rho_2$ and $\text{ref}_0 \to \mu_3$,

$\text{ref}_{\frac{2\pi}{3}} \to \mu_2$, $\text{ref}_{\frac{4\pi}{3}} \to \mu_1$,

then the map σ is one-one and onto. Here the map σ corresponds any $a \in D_3$ to $a\sigma$, which is the corresponding permutation of the vertices of the equilateral triangle given by

Further for any $a, b \in D_3$,

We notice that $(ab)\sigma = (a\sigma)(b\sigma)$, because the effect of first combining any $a, b \in D_3$ and then applying σ comes out to be the same as applying σ to each of a and b and then combining the elements $a\sigma$, $b\sigma$ of S_3. Thus $D_3 \cong S_3$.

Mappings in Group Theory 231

(9) The map $\sigma : (G, \cdot) \to (G, \cdot)$ defined by $a\sigma = a^{-1}$, where G being any group under $* = \cdot$, is an automorphism if G is abelian (see (3) in Illust. Examples 1.9.1)

7.2 Cyclic Groups and Isomorphism

In problem 1.42, we showed that the group of nth roots of unity is isomorphic to the group $(\mathbb{Z}_n, +_n)$. Notice that both the groups are cyclic. Theorem 4.3.9 establishes that any finite cyclic group of order n is isomorphic to the group of residue classes modulo n, $(\mathbb{Z}_n, +n)$. Thus any two cyclic groups of order n are isomorphic. Since we know that all groups of prime orders are cyclic, thus, for example, all groups of order 2 are isomorphic to each other, Similarly all groups of order 3 are isomorphic to each other, and so on. As isomorphic groups are 'copies' of each other, study of any one group of prime order p describes the properties possessed by all the groups of order p.

Non-cyclic groups may also be isomorphic to cyclic groups. Consider the following :

Example 7.2.1 : Klein's 4-group V= {e, a, b, c}, which is a non-cyclic finite group, is isomorphic to the (external) direct product G × G of a cyclic group G = <g> of order 2. The isomorphism is given by the following correspondences :

$$a \to (g, e), b \to (e, g) \text{ and } c \to (g, g).$$

7.3 Properties of Homomorphism

Theorem 7.3.1 : Let $\sigma : G \to G'$ be a *hmp* between two groups (G, *) and (G', 0) and e, e' be respectively the id. elements of G and G'. Then

(i) $e\,\sigma = e'$

(ii) $a^{-1}\sigma = (a\sigma)^{-1}$, where a^{-1} is the inv. of a, and $(ba^{-1})\sigma = (b\sigma)(a\sigma)^{-1}$

(iii) $a^n\,\sigma = (a\sigma)^n$, n is a positive integer

(iv) If $O(a) = m$ then $O(a\sigma)$ is a divisor of m.

Proof : (i) For any $a \in G$, we have
$$e * a = a$$
$$\Rightarrow (e * a)\sigma = a\sigma \quad \text{(as } \sigma \text{ is a well-defined map)}$$
$$\Rightarrow (e\sigma) \, 0 \, (a\sigma) = a\sigma \quad (\sigma \text{ is } hmp)$$
$$= e' \, 0 \, (a\sigma)$$
$$\Rightarrow e\sigma = e' \quad \text{(by RCL)}$$

(ii) Since a^{-1} is inv a
$$\Rightarrow a^{-1} * a = e$$
$$\Rightarrow (a^{-1} * a)\sigma = e\sigma$$
$$= e' \quad \text{(by } i)$$
$$\Rightarrow (a^{-1}\sigma) \, 0 \, (a\sigma) = e' \quad (\sigma \text{ is } hmp)$$
$$\Rightarrow a^{-1}\sigma = (a\sigma)^{-1}$$

Further, for any $a, b \in G$
$$(b * a^{-1})\sigma = (b\sigma) \, 0 \, (a^{-1}\sigma)$$
$$= (b\sigma) \, 0 \, (a\sigma)^{-1},$$
or $\quad (ba^{-1})\sigma = (b\sigma)(a\sigma)^{-1}.$

(iii) $\quad a^n \sigma = (a * a * \ldots a \, (n \text{ times}))\sigma$
$$= (a\sigma) \, 0 \, (a\sigma) \, 0 \ldots 0 \, (a\sigma) \, (n \text{ times})$$
(repeated application of *hmp*-condition)
$$= (a\sigma)^n.$$

(iv) Since $\quad 0(a) = m,$ by (iii)
$$a^m = e \Rightarrow (a^m)\sigma = e\sigma$$
$$= e'$$
$$\Rightarrow (a\sigma)^m = e' \quad \ldots(1)$$

By (1) $O(a\sigma) \leq m$. Let $O(a\sigma) = m_1$ then $(a\sigma)^{m_1} = e'$, which in view of (1) implies that m_1 must be a divisor of m.

Note : By (i), since $e\sigma = e'$ thus atleast $e \in K_\sigma$ so that Kernel K_σ is a non-empty set.

Mappings in Group Theory

Theorem 7.3.2 : Let $\sigma : G \to G'$ and $\tau : G' \to G''$ be two epimorphisms. Then the composite map $\tau\sigma : G \to G''$ is also an epimorphism and K_σ is a subgroup of $K_{\tau\sigma}$.

Proof : For the first part, follow problem 1.39. In order to prove that $K_\sigma < K_{\tau\sigma}$, we need to show that $K_\sigma \subseteq K_{\tau\sigma}$ as K_σ and $K_{\tau\sigma}$ both are normal subgroups of G. Let $a \in K_\sigma$ then $a\sigma = e'$, the id. element of G'. Now $a (\tau\sigma) = (\tau\sigma) a = \tau (\sigma a) = \tau (e') = e''$, where e'' is the id. element of G'' so that $a \in K_{\tau\sigma}$.

Theorem 7.3.3 : The homomorphism $\sigma : G \to G^1$ is monomorphism if and only if $K_\sigma = \{e\}$, where e is the identity element of G.

Proof : (i) *First part :* Assume that σ is a monomorphism. Then by definition,

$$K_\sigma = \{x \in G : x\sigma = e'\} = \{x \in G : x\sigma = e\sigma\} = \{x \in G : x = e\}$$

(as σ is one-one)

$\Rightarrow K\sigma = \{e\}$

(ii) *Second part :* Assume that $K_\sigma = e\}$

Then $x \sigma = y\sigma$, where $x, y \in G$,

$\Rightarrow (x \sigma)(y\sigma)^{-1} = e^1 \Rightarrow (x\sigma)(y^{-1}\sigma) = e' \Rightarrow (xy^{-1}) \sigma = e'$

$\Rightarrow \qquad xy^{-1} \in K_\sigma$

But $\qquad K_\sigma = \{e\}$, thus $xy^{-1} = e$

$\Rightarrow \qquad x = y$

i.e. $\qquad x\sigma = y\sigma$

$\Rightarrow \qquad x = y$

or σ is one-one (monomorphism).

Theorem 7.3.4 : (Correspondence theorem)

Let $\sigma : G \to G'$ be a homomorphism. Then,

(i) $H < G \Rightarrow (H) \sigma < (G) \sigma$

(ii) $H \triangleleft G \Rightarrow (H) \sigma \triangleleft (G) \sigma$ and

(iii) $H^1 < (G) \sigma \Rightarrow (H^1) \sigma^{-1} < G$

(iv) $H^1 \triangleleft (G) \sigma \Rightarrow (H^1) \sigma^{-1} \triangleleft G$

(If σ is onto as well, in all the above results (G) σ = G′)

Proof: (i) (H) σ = {hσ : h ∈ H}. In order to prove that (H) σ < (G) σ, where (H) σ is a non-empty subset of (G) σ, take two elements h_1^1 and h_2^1 of (H) σ

Then

$h_1' = h_1\sigma$ and $h_2' = h_2\sigma$, for some $h_1, h_2 \in H$, and

$h_1' h_2'^{-1}$

$= (h_1\sigma)(h_2\sigma)^{-1}$

$= (h_1\sigma)(h_2^{-1}\sigma)$

$= (h_1 h_2^{-1})\sigma$ ...(1) (as σ is *hmp*)

Since H < G, thus $h_1 h_2^{-1} \in H$ so that $(h^1 h_2^{-1})\sigma \in (H)\sigma$, or by (1), $h_1' h_2'^{-1} \in (H)\sigma$ whenever $h_1', h_2' \in (H)\sigma$ i.e., (H) σ < (G) σ.

(ii) By (i), (H) σ < (G) σ. We need to prove that (H) σ is normal as well. Let a′ ∈ (G) σ and h′ ∈ (H) σ. We will prove that $a' h' a'^{-1} \in (H)\sigma$.

a′ ∈ (G)σ ⇒ a′ = aσ for some a ∈ G, and h′ ∈ (H) σ ⇒ h′ = hσ for some h ∈ H then $a' h' a'^{-1}$

$= (a\sigma)(h\sigma)(a\sigma)^{-1}$

$= (a\sigma)(h\sigma)(a^{-1}\sigma)$ (σ is *hmp*)

$= (aha^{-1})\sigma$ ...(2) (*hmp*-condition)

Since H ◁ G thus $aha^{-1} \in H$ so that $(aha^{-1})\sigma \in (H)\sigma$, or by (2), $a'h'a'^{-1} \in (H)\sigma$.

(iii) Let $h_1, h_2 \in (H')\sigma^{-1}$. We will prove that $h_1 h_2^{-1} \in (H')\sigma^{-1}$.

Mappings in Group Theory

$h_1 \in (H')\sigma^{-1} \Rightarrow h_1\sigma \in H_1$, and
$h_2 \in (H')\sigma^{-1} \Rightarrow h_2\sigma \in H_1$
Since $H' < (G)\sigma$

$\Rightarrow \quad (h_1\sigma)(h_2\sigma)^{-1} \in H'$

$\Rightarrow \quad (h_1\sigma)(h_2^{-1}\sigma) \in H'$

$\Rightarrow \quad (h_1 h_2^{-1})\sigma \in H'$

$\Rightarrow \quad h_1 h_2^{-1} \in (H')\sigma^{-1}.$

(iv) By (iii) $(H')\sigma^{-1} < G$. We need to prove that $(H')\sigma^{-1}$ is normal as well. Let $a \in G$ and $h \in (H')\sigma^{-1}$. We will prove that $aha^{-1} \in (H')\sigma^{-1}$.

$$a \in G \Rightarrow a\sigma \in (G)\sigma, \text{ and}$$
$$h \in (H')\sigma^{-1} \Rightarrow h\sigma \in H'$$

Then, since $H^1 \triangleleft (G)\sigma$,

$\quad\quad (a\sigma)(h\sigma)(a\sigma)^{-1} \in H'$

$\Rightarrow \quad (a\sigma)(h\sigma)(a^{-1}\sigma) \in H'$

$\Rightarrow \quad (aha^{-1})\sigma \in H'$

$\Rightarrow \quad aha^{-1} \in H'\sigma^{-1}$

Theorem 7.3.5 : Kernel K_σ of a homomorphism $\sigma : G \to G'$ is normal subgroup of G.

Proof : By definition

$K_\sigma = \{x \in G : x\sigma = e'\}$

and K_σ iss a non-empty subset of G.

(i) K_σ is a subgroup of G : Let $h, k \in K_\sigma$. We will prove that $h\,k^{-1} \in K_\sigma$

$h \in K_\sigma \Rightarrow h\sigma = e'$, and

$k \in K_\sigma \Rightarrow k\sigma = e'$

Then $(hk^{-1})\sigma$

$= (h\sigma)(k^{-1}\sigma) = (h\sigma)(k\sigma)^{-1} = (e')(e')^{-1} = e'$

or, $hk^{-1} \in K_\sigma$.

(ii) K_σ is normal : Let $a \in G$ and $k \in K_\sigma$. We will prove that $a k a^{-1} \in K_\sigma$ or $(a k a^{-1}) \sigma = e'$.

Consider

$(aka^{-1})\sigma$

$= (a\sigma)(k\sigma)(a^{-1}\sigma) = (a\sigma)(e')(a\sigma)^{-1} = (a\sigma)(a\sigma)^{-1} = e'$

(as $K \in K_\sigma$, $K\sigma = e'$)

Theorem 7.3.6 : Let N be a normal subgroup of a group G so that $\dfrac{G}{N}$ is a factor group. Then the map $v : G \to \dfrac{G}{N}$ defined by

$av = aN$ for all $a \in G$ is an epimorphism and $N = K_v$.

Proof: Take an arbitrary element xN of $\dfrac{G}{N}$, then $x \in G$ and $x v = x N$ so that v is an onto map. Further for any $a, b \in G$,

$(ab)v = ab\ N$

$= (aN)(bN)$

$= (av)(bv)$,

thus v is a *hmp*.

(Notice that v is not one-one). Now, in order to prove that $K_v = N$, we will show that $K_v \subseteq N$ and $N \subseteq K_v$.

Since $v : G \to \dfrac{G}{N}$ is defined as $xv = xN$ for any $x \in G$; if $x \in K_v$ we have $xv =$ id. element of $\dfrac{G}{N}$ (by definition of K_v) $= N$

Thus $xN = N \Rightarrow x \in N$, i.e., $x \in K_v \Rightarrow x \in N$ or $K_v \subseteq N$. Next, if $n \in N$, $nv = nN$ ($N \subseteq G$)

$= N$ ($nN = N$ if $n \in N$)

i.e., $nv = N \Rightarrow n \in K_v$ or $N \subseteq K_v$.

Mappings in Group Theory 237

Note : (1) We can define $v : G \to \frac{G}{N}$ as $av = Na$ for all $a \in G$.

(2) The map $v : G \to \frac{G}{N}$, where $N = K_v$, is called *natural* or *canonical homomorphism*.

(3) Above theorem also tells that every normal subgroup of G is Kernel of some *hmp*.

The following theorem establishes an isomorphism between factor group $\frac{G}{N}$ and the homomorphic image G' of the homomorphism $\sigma : G \to G'$, where $N = K_\sigma$.

Theorem 7.3.7 *(Fundamental homomorphism theorem)* : Let $\sigma : G \to G'$ be an epimorphism and K_σ be its kernel then $\frac{G}{K_\sigma} \cong G'$.

Proof : Since kernel K_σ is a normal subgroup of G, $\frac{G}{K_\sigma}$ is a valid factor group. We will prove that there exists an isomorphic map ϕ between $\frac{G}{K_\sigma}$ and the homomorphic image G^1 of the homomorphism $\sigma : G \to G'$.

Let the map $\phi : \frac{G}{K_\sigma} \to G'$ be defined by

$$(aK_\sigma)\phi = a\sigma \quad \text{for all } aK_\sigma \in \frac{G}{K_\sigma}. \qquad ...(1)$$

Here $a\sigma \in G'$ and since σ is onto, $a \in G$.

(*i*) Map ϕ is well-defined :

For $a, b \in K_\sigma$, let $aK_\sigma = bK_\sigma$

$\Rightarrow \quad ab^{-1} \in K_\sigma$

$\Rightarrow \quad (ab^{-1})\sigma = e'$, where e' is identity element of G'.

$\Rightarrow \quad (a\sigma)(b^{-1}\sigma) = e'$

$\Rightarrow \quad (a\sigma)(b\sigma)^{-1} = e'$

$\Rightarrow \quad (a\sigma) = (b\sigma)$

$\Rightarrow \quad (aK_\sigma)\phi = (bK_\sigma)\phi \qquad$ (by 1)

Thus, $aK_\sigma = bK_\sigma \Rightarrow (aK_\sigma)\phi = (bK_\sigma)\phi$,

i.e., ϕ is a map.

(ii) ϕ is one-one : We will prove that kernel ϕ, $K_\phi = \{K_\sigma\}$. where K_σ is the id. element of the group $\dfrac{G}{K_\sigma}$

Let $\quad aK_\sigma \in K_\phi$

$\Rightarrow \quad (aK_\sigma)\phi = e'$

$\Rightarrow \quad a\sigma = e' \qquad\qquad\qquad\qquad$ (by 1)

$\Rightarrow \quad a \in K_\sigma$

$\Rightarrow \quad aK_\sigma = K_\sigma$ (If $H < G$ and $h \in H$ then $hH = H$)

Thus $aK_\sigma \in K_\phi \Rightarrow aK_\sigma = K_\sigma$, i.e., $K_\phi = \{K_\sigma\}$

(iii) ϕ is onto : For any $a' \in G'$,

we have $a \in G$ such that

$a' = (a)\sigma \quad (\sigma$ is onto)

Thus

$(aK_\sigma)\phi$ which by (1) is $a\sigma$ has value a' in G'.

(iv) ϕ is *hmp* : we will prove that

$\{(aK_\sigma)(bK_\sigma)\}\phi = (aK_\sigma)\phi\,(bK_\sigma)\phi$,

where $aK_\sigma, bK_\sigma \in \dfrac{G}{K_\sigma}$

LHS, by definition of product of two left cosets, equals $(abK_\sigma)\phi$

$\qquad\qquad\qquad = (ab)\sigma \qquad$ (by 1)

Mappings in Group Theory 239

$$= (a\sigma)(b\sigma) \quad (\sigma \text{ is } hmp)$$
$$= (aK_\sigma)\phi (bK_\sigma)\phi \quad \text{(by 1)}$$

which is the RHS. Thus $\phi : \frac{G}{K_\sigma} \to G'$ is an isomorphism, so that $\frac{G}{K_\sigma} \cong G'$.

Note : (1) The situations in Theorem 7.3.6 and 7.3.7 can be described by the following diagram

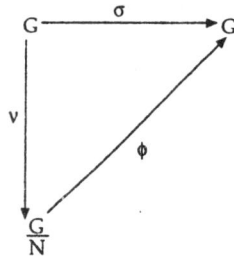

The map $\sigma : G \to G'$ is an epimorphism, the map $v : G \to \frac{G}{N}$ is also an epimorphism (called the natural *hmp*) and the map $\phi : \frac{G}{N} \to G^1$ is an isomorphism. For v, N in $\frac{G}{N}$ is K_v and for ϕ, N in $\frac{G}{N}$ is K_σ.

(2) Notice that $\sigma = \phi\, 0v = \phi v$

Since $(\phi v)\, a$, where $a \in G$,
$= \phi(va) = \phi(av) = \phi(aN) = (aN)\phi = a\sigma$
or, $a(\phi v) = a\sigma$, for all $a \in G$.

Theorem 7.3.8 : Let $\sigma : G \to G'$ be a homomorphism with Kernel K_σ and N be a normal subgroup of G such that $N \subseteq K_\sigma$. Then there is a unique homomorphism $\tau : \frac{G}{N} \to G'$ such

that $(aN)\tau = a\sigma$ for all $a \in G$ and image of τ = Image of σ and
$$K_\tau = \frac{K_\sigma}{N}$$

Further τ is an isomorphism if and only if σ is an epimorphism and $K_\sigma = N$.

Proof: Given: (i) $\sigma : G \to G^1$ is a *hmp* with Kernel K_σ and $N \triangleleft G$

(ii) $N \subseteq K_\sigma$.

we will prove that following:

(i) There exists a unique *hmp* $\tau : \frac{G}{N} \to G'$ defined by $(aN)\tau = a\sigma$, for all $a \in G$, and $a\sigma \in G'$.

(ii) Image of τ, $\left(\frac{G}{N}\right)\tau$ = Image of σ, $(G)\sigma$

(iii) Kernel of τ, $K_\tau = \frac{K_\sigma}{N}$

(iv) τ is an isomorphism $\Leftrightarrow$ σ is onto and $K\sigma = N$.

By definition, $\frac{G}{N} = \{aN, bN, cN, \ldots : a, b, c, \ldots \in G\}$

Also,
$$G' = \{a\sigma, b\sigma, c\sigma, \ldots : a, b, c, \ldots \in G\}$$

Let $t \in aN$. Then there is some n in N such that $t = an$ and $(t)\sigma = (an)\sigma = (a\sigma)(n\sigma) = a\sigma$

(as σ is *hmp* and $n \in N \subseteq K_\sigma$).

Thus $(t)\sigma = a\sigma$. where t is an arbitrary element of aN, so that the map $\tau : \frac{G}{N} \to G'$ defined by $(aN)\tau = a\sigma$ is well defined.

Further, for any aN, bN in $\frac{G}{N}$, we have $\{(aN)(bN)\}\tau =$
$\{(abN)\tau = (abN)\sigma = (a\sigma)(b\sigma) = (aN)\tau\, (bN)\tau$ thus τ is a *hmp*.

Since $\left(\frac{G}{N}\right)\tau = \{(aN)\tau, (bN)\tau, (cN)\tau, \ldots : a, b, c, \ldots \in G\}$

$\qquad\qquad = \{a\sigma, b\sigma, c\sigma, \ldots : a, b, c, \ldots \in G\}$

and $\qquad (G)\,\sigma = \{a\sigma, b\sigma, c\sigma, \ldots : a, b, c, \ldots \in G\}$

thus $\left(\frac{G}{N}\right)\tau = (G)\,\sigma$, i.e.,

Image of τ = Image of σ

Also

$aN \in K_\tau \Leftrightarrow (aN)\tau$ i.e., $a\sigma = e'$, where e' is id. element of G'. as $= e'$ implies $a \in K\sigma$

Thus $\qquad K_\tau = \{aN : a \in K\sigma\}$

$\qquad\qquad = \frac{K_\sigma}{N}$

From above it follows that the map τ is a monomorphism if and only if K_τ is an improper subgroup of $\frac{G}{N}$ which holds if and only if $K_\tau = N$. Thus τ is a monomorphism $\Leftrightarrow K\sigma = N$. Further, by definition, τ is an epimorphism if and only if σ is an epimorphism, from which we conclude that τ is an isomorphism $\Leftrightarrow$ (i) $K_\sigma = N$ and (ii) σ is an epimorphism where

$\tau : \frac{G}{N} \to G'$ and $\sigma : G \to G'$.

Since σ completely determines τ, the map τ is unique.

From this theorem, we have following two important deductions :

Corollary 7.3.1 : (First Isomorphism theorem) The homomorphism $\sigma : G \to G'$ between two groups G and G' induces an isomorphism τ between the groups $\dfrac{G}{K_\sigma}$ and $(G)\sigma$

Corollary 7.3.2 : (Fundamental *hmp* theorem) : If $\sigma : G \to G'$ is an epimorphism then $\dfrac{G}{K_\sigma} \cong G'$, which is our Theorem 7.3.7.

7.4 Isomorphism Theorems and Cayley's Theorem

Theorem 7.4.1 : (Second Isomorphism theorem) If N and K are any two normal subgroups of a group G such that $K \subseteq N$ then $G|N \cong (G|K) | (N|K)$.

Proof : We know that $G|K$, $G|N$ and $N|K$ are groups and $N|K$ is a normal subgroup of $G|K$ (Prob 6.10) Define a map $\sigma : G|K \to G|N$ by $(aK) \sigma = aN$ for all $a \in G$.

Now for any $a, b, \in G$,

$$aK = bK \Rightarrow a^{-1} b \in K \Rightarrow a^{-1} b \in N \text{ (as } K \subseteq N)$$
$$\Rightarrow aN = bN$$
$$\Rightarrow (aK)\sigma = (bK)\sigma,$$

implying that σ is a well-defined map. We will prove that σ is an epimorphism and then fundamental *hmp* theorem when applied on the map σ will give us the result.

Since

$$\{(aK)(bK)\} \sigma = \{(ab) K\} \sigma = ab\, N$$
$$= (aN)(bN)$$
$$= (aK)\sigma\, (bK)\sigma,$$

implying that σ is a *hmp*.

Further, for any $aN \in G|N$, where $a \in G$, there is $aK \in G|K$ satisfying $(aK) \sigma = aN$ so that σ is onto.

Thus $\sigma : G|K \to G|N$ is an onto *hmp*.

Finally Kernel of σ,
$$K\sigma = \{aK \in G|K : (aK)\sigma = aN = N\}$$
$$= \{aK : a \in N\} \quad (\text{as } aN = N \Rightarrow a \in N)$$
$$= N/K$$

Thus by fundamental *hmp* theorem
$$G|N \cong (G|K) | (N|K).$$

Theorem 7.4.2. (Third iomorphism theorem) If H is a subgroup and N is a normal subgroup of a group G then
$$HN|N > H|H \cap N$$

Proof : We know that if $H < G$ and $N \triangleleft G$ then $H \cap N \triangleleft H$. This $H|H \cap N$ is a valid factor group. We also know that $HN < G$. Since $N \triangleleft G$ and $N \subseteq HN$, $N \triangleleft HN$. Thus $HN|N$ is a valid factor group. Define a map $\sigma : H \to HN|N$ by $h\sigma = hN$ for all $h \in H$.

The map σ is well defined as $H \subseteq HN$ and this $h \in H \Rightarrow h \in HN$ so that $h N \in HN|N$. We will prove that σ is an epimorphism and then fundamental *hmp* theorem when applied on the map σ will give us the result.

Since for any $h_1, h_2 \in H$,

$(h_1 h_2)\sigma = (h_1 h_2) N = (h_1 N)(h_2 N) = (h_1\sigma)(h_2\sigma)$, implying that σ is a *hmp*.

In order to prove that σ is onto, we proceed as follows :

Take an arbitrary element hN of $HN|N$.

Then $h \in HN \Rightarrow$ there exists $h_1 \in H$ and $n_1 \in N$ such that
$$h_1 = h_1 n_1 \quad ...(1)$$
Since $\quad N \triangleleft G, HN = NH \quad$ for some $n_2 \in N$ and
$\Rightarrow \qquad\qquad h_1 n_1 = n_2 h_2, - (2)\ h_2 \in H.$
Now

$$h_2 \sigma = h_2 N$$
$$= h_2 (n_2 N)(\text{as } n_2 \in N \Rightarrow n_2 N = N)$$
$$= (h_2 n_2) N$$

$$= (h_1\, n_1)\, N \qquad \text{(by 2)}$$
$$= hN \qquad \text{(by 1)}$$

Thus $hN \in HN|N \Rightarrow$ there is some $h_2 \in H$ such that $h_2 \sigma$ hN, implying that σ is onto.

Finally Kernel of σ,

$$K_\sigma = \{h \in H : h\,\sigma = hN = N\}$$
$$= \{h \in H : h \in N\} \text{ (as } hN = N \Rightarrow h \in N)$$
$$= \{h : h \in H \text{ and } h \in N\}$$
$$= H \cap N$$

Thus by fundamental *hmp* theorem,

$$HN|N \cong H|H \cap N$$

Cayley was the first to observe that all groups are isomorphic to subgroups of some transformation groups. Cayley's results are contained in the following theorem (bearing his name) and its corollary.

Theorem 7.4.3 (Cayley's theorem) : Any group G is isomorphic to subgroup of a transformation group.

Proof : The set of all one-to-one maps from G onto itself forms a group B_G under composition of maps and is called the transformation group of G (see problem 1.19). For each element a in G, the map defined by $x\,\sigma_a = ax, x \in G$, is one-to-one onto because

(i) $x\sigma_a = y\sigma_a \Rightarrow ax = ay \Rightarrow x = y$ (by LCL), and, for $b \in G$,

(ii) Since $(a^{-1}b)\,\sigma_a = a\,(a^{-1}b) = b$ implying that $a^{-1}b \to b$.

Thus $\sigma_a \in B_G$. Consider set J of all such maps defined for each element of G, i.e., $J = \{\sigma_a : a \in G\}$,

then J is a subset of B_G. Our next objective is to prove that :

J is a subgroup of B_G and $G \cong J$. Let $\sigma_a, \sigma_b \in J$ where $a, b \in G$. Then, in the notation of composite map [see

Mappings in Group Theory

Prelim., 3 (vi)], $(\sigma_a \sigma_b) x = \sigma_a (\sigma_b x) = \sigma_a (bx) = a (bx) = (ab) x = \sigma_{ab} x$

or,

$\sigma_a \sigma_b = \sigma_{ab}$. Since $ab \in G$, $\sigma_{ab} \in J$ so that $\sigma_a, \sigma_b \in J \Rightarrow \sigma_a \sigma_b \in J$.

Further,

$(\sigma_a \sigma_{a^{-1}}) x = \sigma_a (\sigma_{a^{-1}} x) = \sigma_a (a^{-1} x) = a (a^{-1} x)$
$= (a\ a^{-1}) x = x = Ix,$

where I is the identity map in B_G.

Thus $\sigma_a \sigma_{a^{-1}} = I$, or, $(\sigma_a)^{-1} = \sigma_{a^{-1}}$. But $\sigma_{a^{-1}} \in J$, thus $(\sigma_a)^{-1} \in J$ so that $\sigma_a \in J \Rightarrow (\sigma_a)^{-1} \in J$. Hence J is a subgroup of B_G.

Finally, the map $f : G \to J$ defined by $af = \sigma_a$, for all $a \in G$, is an isomorphism because

(iii) $af = bf \Rightarrow \sigma_a = \sigma_b \Rightarrow x \sigma_a = x\sigma_b, \quad x \in G$,
i.e., f is one-to-one, $\Rightarrow ax = bx \Rightarrow a = b$ (by LCL),

(iv) $(ab) f = \sigma_{ab} = \sigma_a \sigma_b = (af) (bf)$, for any $a, b \in G$, i.e., f is hmp, and

(v) by definition for any $\sigma_x \in J$, $x \in G$, i.e., f is onto.

Thus $G \cong J$, i.e., any group G is isomorphic to subgroup J of the transformation group B_G.

Arthur Cayley (1821-1895) : Born on August 16, 1821 Cayley started his career as a lawyer in 1849, and in 1863 he became Professor of Pure Mathematics at Cambridge. Cayley worked in the fields of abstract algebra, astronomy, theoretical dynamics and matrix theory and published 966 research papers on these topics. The axiomatic definition of a group, introduction of composition table for a finite group (Cayley's table) and formulation of problem of finding all finite groups of order n are his significant contribution in the field of theory of groups. He presented the famous theorem of matrix – Cayley-Hamilton theorem – jointly with Sir Willian R Hamilton (1805-1865).

Corollary 7.4.1 : Every finite group of order n is isomorphic to a subgroup of the group S_n.

Proof : The proof is exactly on the lines of proof of above theorem. Here $B_G = S_n$ and while proving that the subset J of S_n is its subgroup, we need only to show the closure property in J, as S_n is a finite group.

Note : (1) The subgroup J of S_n is said to be a *regular permutation group*. For example, in problem 4.6, the subgroup $\{I, \alpha_1, \alpha_2, \alpha_3\}$ of group S_4 is a regular permutation group.

(2) The group D_4 is not isomorphic to S_4, (note that $D_3 \cong S_3$) as their orders are different, but D_4 is isomorphic to subgroups of S_4. Find out one such subgroup of S_4.)

7.5 Group of Automorphisms

Theorem 7.5.1 : The set of all automorphisms of a group G, denoted by Aut G, forms a group under composition of maps.

Proof : Here Aut G = $\{\sigma : \sigma$ is an automorphism of G$\}$

so that $\sigma : G \to G$ is an isomorphism and $*$ = composition of maps.

(i) Let $\sigma, \tau \in$ Aut G then, as shown in problem 1.39, $\sigma\tau : G \to G$ is an isomorphism so that $\sigma \tau \in$ Aut G.

(ii) Composite of arbitrary maps follows A.L. (see Prelim. 3 (*vi*))

(iii) The identity map I on G, defined by $aI = a$ for all $a \in G$, satisfies $I\sigma = \sigma I = \sigma$ for any $\sigma \in$ Aut G so that I is the id. element for $*$ over Aut G. Since I is an isomorphism and for any $a, b \in G$,

$$(ab) \, I = ab = (a) \, I \, (b) \, I,$$

thus $I \in$ Aut G,

Mappings in Group Theory 247

(iv) For any $\sigma \in$ Aut G, σ^{-1} exists and it is one-one onto as σ is an one-one onto map. Thus $\sigma^{-1} : G \to G$ is an isomorphism and for any $a, b \in$ co-domain G, there exists a_1, b_1 in the domain G such that

$$a\sigma^{-1} = a_1 \text{ and } b\sigma^{-1} = b_1, \text{ or } a_1 \sigma = a \text{ and}$$
$$b_1\sigma = b \text{ and we have}$$
$$(ab) \sigma^{-1} = \{(a_1 \sigma) (b_1 \sigma)\} \sigma^{-1} = \{(a_1 b_1) \sigma\} \sigma^{-1} = a_1 b_1$$
$$= (a\sigma^{-1}) (b\sigma^{-1})$$

or σ^{-1} is a *hmp*. Thus if $\sigma \in$ Aut G, $\sigma^{-1} \in$ Aut G so that each element in Aut G has its inverse in Aut G.

Theorem 7.5.2 : For a fixed element a of a group G, the map

$$i_a : G \to G \text{ defined by}$$
$$b_{i_a} = aba^{-1}$$

is an automorphism of G.

Proof : We will show that the map i_a for a fixed element a of G is one-one *hmp* of G onto G.

(i) For any $b, c \in$ G

$$b \; i_a = c \; i_a \Rightarrow a \; b \; a^{-1} = a \; c \; a^{-1}$$
$$\Rightarrow a^{-1} (a \; b \; a^{-1}) a = a^{-1} (a \; c \; a^{-1}) a \quad \text{(by LCL and RCL)}$$
$$\Rightarrow \qquad b = c \qquad\qquad\qquad \text{(by AL)},$$

or i_a is one-to-one.

(ii) Let for any y in co-domain G,

$$x \; i_a = y$$
$$\Rightarrow \qquad a \; x \; a^{-1} = y$$
$$\Rightarrow \qquad ax = ya$$
$$\Rightarrow \qquad x = a^{-1} ya, \text{ then } x \text{ is in domain G}$$
$$\Rightarrow \qquad a^{-1} ya \to y \text{ or } i_a \text{ is onto.}$$

(iii) Finally, for any $b, c \in$ G

$$(bc) \; i_a = a \; (bc) \; a^{-1}$$

$$= (a\ b\ a^{-1})\ (a\ c\ a^{-1})$$
$$= (b\ i_a)\ (c\ i_a)$$

i.e., i_a is a *hmp*. Thus $i_a \in$ Aut G.

The map i_a is called an *inner automorphism determined by the element a*.

Definition 7.5.1 : The set of all inner automorphism i_a for all $a \in$ G is the set $I_N(G) = \{i_a : i_a$ is inner automorphism and a is any element of G$\}$

Theorem 7.5.3 : Let G be a group. Then $I_N(G) \triangleleft$ Aut G.

Proof : For any $y_{ia} \in I_N(G)$, we have
$$bi_a = a\ b\ a^{-1} \text{ where } b \in G.$$

Now, following the notation of composite maps,
$$(i_a\ i_{a^{-1}})\ b = i_a\ (i_{a^{-1}}\ b) = i_a\ (a^{-1}\ b\ (a^{-1})^{-1})$$
$$= i_a\ (a^{-1}\ ba)$$
$$= a\ (a^{-1}\ ba)a^{-1}$$
$$= b$$

$\Rightarrow i_a\ i_{a^{-1}}$ is the identity map on G,

i.e., $\qquad i_{a^{-1}} = (i_a)^{-1} \qquad$ ...(1)

(i) To prove that $I_N(G) <$ Aut G

For any i_a and i_b in $I_N(G)$, and $x \in G$ $\{ia\ (i_b)^{-1}\ x = \{i_a\ (i_{b^{-1}})\}\ x$
(by 1)
$$= i_a\ (i_{b^{-1}}\ x) = i_a\ (b^{-1}\ x\ (b^{-1})^{-1})$$
$$= i_a\ (b^{-1}\ xb) = a\ (b^{-1}\ x\ b)\ a^{-1}$$
$$= (ab^{-1})\ x\ (ba^{-1}) = (ab^{-1})\ x\ (ab^{-1})-1 = i_{ab^{-1}}\ x$$

Since $ab^{-1} \in G$, $i_{ab^{-1}} \in I_N(G)$

Mappings in Group Theory

Thus $i_a \in G$, $i_b \in I_N(G) \Rightarrow (i_a)(i_b)^{-1} \in I_N(G)$
or $I_N(G) < \text{Aut } G$

(ii) For any $\sigma \in \text{Aut } G$, consider
$\sigma^{-1} i_a \sigma$ where $i_a \in I_N(G)$
For $x \in G$,

$$\begin{aligned}(\sigma^{-1} i_a \sigma) x &= \sigma^{-1} i_a \sigma(x) \\ &= \sigma^{-1}(a \sigma(x) a^{-1}) \quad \text{(by definition of } i_a) \\ &= \sigma^{-1}(a t a^{-1}) \quad (\sigma(x) = t) \\ &= \sigma^{-1}(a) \sigma^{-1}(t) \sigma^{-1}(a^{-1}) \quad (\sigma^{-1} \text{ is } hmp) \\ &= \sigma^{-1}(a) x \sigma^{-1}(a^{-1}) = \sigma^{-1}(a) x (\sigma^{-1} a)^{-1} \\ &= y x y^{-1} \quad (\sigma^{-1}(a) = y \in G) \\ &= i_y x\end{aligned}$$

Since $i_y \in I_N(G)$, therefore $\sigma^{-1} i_a \sigma \in I_N(G)$
i.e., $I_N(G) \triangleleft \text{Aut } G$.

(Aut $G | I_N(G)$ is an example of a factor group.)

Theorem 7.5.4: If $Z(G)$ is the centre of a group G then $I_N(G) \cong G | Z(G)$.

Proof: We will apply fundamental hmp theorem to prove this result. Define a map $f : G \to I_N(G)$ by the rule $af = \sigma_a$ for all $a \in G$. We will prove the following :

(i) f is an epimorphism, and

(ii) Kernel of f, $K_f = Z(G)$.

For any $i_x \in I_N(G)$, $x \in G$ so that
$x \to i_x$ or f is onto.
For any $x, y \in G$ $(xy)f = \sigma_{xy} = \sigma_x \sigma_y = (xf)(yf)$
i.e., f is hmp. Thus the map f is an epimorphism.
Let $x \in G$ and $x \in K_f$, then
$xf = i_x = I_G$ where I_G is the identity function on G and
hence I_G is the identity element of $I_N(G)$.

Now $\qquad i_x = I_G$
$\Rightarrow \qquad y\, i_x = y\, I_G$ for all $y \in G$
$\Rightarrow \quad x y x^{-1} = y \Rightarrow xy = yx \Rightarrow x \in Z(G)$
Thus $\quad x \in K_f \Rightarrow x \in Z(G)$, or $K_f \subseteq Z(G)$.
Next, if $x \in Z(G)$, then $xy = yx$ for all $y \in G$
$\Rightarrow \qquad xy\, x^{-1} = y$
$\Rightarrow \qquad y\, i_x = y\, I_G \Rightarrow i_x = I_G \Rightarrow xf = I_G$
$\Rightarrow \qquad x \in K_f$

Thus $x \in Z(G) \Rightarrow x \in K_f$ or $Z(G) \subseteq K_f$, so that $K_f = Z(G)$.
By fundamental theorem applied on $f : G \to I_N(G)$, we have

$$I_N(G) \cong G\,|\,K_f, \text{ or}$$
$$I_N(G) \cong G\,|\,Z(G).$$

Solved Problems Set

Problem 7.1 : Let G and G' be groups and $\sigma : G \to G'$ be an epimorphism. Prove that if G is abelian then G' is abelian.

Solution : Let $a^1, b^1 \in G'$. Since σ is onto, there are $a, b \in G$ such that

$$a^1 = a\sigma \text{ and } b^1 = b\sigma, \text{ so that}$$
$$a^1 b^1 = (a\sigma)(b\sigma) = (ab)\,\sigma \text{ (as } \sigma \text{ is } hmp)$$
$$= (ba)\,\sigma \qquad \text{(as G is abelian)}$$
$$= (b\sigma)(a\sigma)$$
$$= b^1 a^1$$

showing that G' is also abelian.

Problem 7.2 : Let $\sigma : G \to G^1$ be an isomorphism. If $a \in G$ and $O(a) = n$ then prove that $O(a\sigma) = n$, where $a\sigma \in G^1$.

Solution : Here $a^n = e$. Since $\sigma : G \to G^1$ is an isomorphism, so is σ^{-1}. Now, $(a\sigma)^n = a^n\,\sigma$

$= e\,\sigma = e^1$, where e^1 is id. element of G^1, i.e.,

$(a\sigma)^n = e^1 \Rightarrow O(a\sigma)$ is finite and $\leq n$.

Let $O(a\sigma) = m$, then $m \leq n$ ...(1)

Exactly proceeding with σ^{-1} and $a\sigma$ in place of σ and a respectively we get $n \leq m$...(2)

By (1) and (2), $n = m$, i.e., $O(a) = O(a\sigma)$.

(*Note* : This result can be used to check if the two groups are in isomorphism.)

Problem 7.3 : Let $\sigma : G \to G^1$ be an isomorphism. Prove that if G is cyclic then G^1 is cyclic.

Solution : Let $G = <a>$ and $a\sigma = a^1$ so that $a^1 \in G^1$. Take an arbitrary element b^1 of G^1 then $b^1 = (a^m)\,\sigma$, for some integer m

$$= (a\sigma)^m$$
$$= (a^1)^m$$

$\Rightarrow G^1 = <a^1>$ or G^1 is cyclic generated by its element a^1.

Problem 7.4 : Let G be a group and G^1 be a groupoid (a non-empty set equipped with a b.o.) and $\sigma : G \to G^1$ be one-to-one onto map satisfying $(ab)\,\sigma = (a\sigma)(b\sigma)$ for all $a, b \in G$ then prove that G^1 is also a group.

Solution : We have to check all the group conditions for given groupoid G^1.

(*i*) A.L. : Take any three elements a^1, b^1, c^1 form G^1. Since map σ is onto, there are a, b, c in G such that $a^1 = a\sigma$, $b^1 = b\sigma$ and $c^1 = c\sigma$. Then

$$a^1\,(b^1\,c^1)$$
$$= (a\sigma)\,[(b\sigma)\,(c\sigma)] = (a\sigma)\,[(bc)\,\sigma] = [a\,(bc)]\,\sigma$$
$$= [(ab)\,c\,\sigma \qquad \text{(as G is a group)}$$
$$= [(ab)\,\sigma]\,[c\,\sigma] = [(a\sigma)\,(b\sigma)]\,(c\sigma) = (a^1 b^1)\,c^1$$

Thus,

$a^1\,(b^1 c^1) = (a^1\,b^1)\,c^1$, or elements in G^1 obey A.L.

(*ii*) Existence of id. element in G^1 :

Since $e \in G$, $e\,\sigma \in G^1$. Then $a^1(e\sigma) = (a\sigma) = (ae)\,\sigma = a\,\sigma$
$= a^1$, i.e., $a^1\,(e\sigma) = a^1$; and

$(e\sigma) a^1 = (e\sigma)(a\sigma) = (ea)\sigma = a\sigma = a^1$,

i.e., $(e\sigma) a^1 = a^1$ so that $e\sigma$ is the identity element in G1.

(iii) *Existence of inverses in* G^1 :

If $a \in G$ then $a^{-1} \in G$, $(a^{-1})\sigma \in G^1$

Then

$a^1 (a^{-1}\sigma) = (a\sigma)(a^{-1}\sigma) = (a^{-1}a)\sigma = e\sigma$, which is id. element of G^1.

Similarly $(a^{-1}\sigma)a^1 = (a^{-1}\sigma)(a\sigma) = (a^{-1}a)s = e\sigma$ showing that $(a^{-1}\sigma) =$ Inv. of a^1. Thus each element if G^1 has its inverse in G^1.

From (i) – (iii), G^1 is a group, isomorphic to G.

Problem 7.5 : If a map $\sigma : G \to G$, where G is a group, is defined by $a\sigma = a^m$ (where m is a fixed integer) for all $a \in G$ then prove that σ is an endomorphism if G is abelian.

Solution : For any $a, b \in G$,

$$(ab)\sigma = (ab)^m \text{ (as G is abelian)}$$
$$= a^m b^m$$
$$= (a\sigma)(b\sigma),$$

so that σ is an endomorphism.

Problem 7.6 : If $\sigma : (\mathbb{R}, +) \to (\mathbb{C}^*, \cdot)$ be a homomorphism defined by $x\sigma = e^{2\pi i x}$ for all $x \in \mathbb{R}$,

then prove that

$$O_1 \cong \mathbb{R} \mid \mathbb{Z}$$

where $(O_1, \cdot)$ is the group of all complex numbers with modules 1, *i.e.*, the circle group (see A 5 (iii) of Illus. Examples 1.4.1) and $(\mathbb{Z}, +)$ is the group of all integers.

Solution : Since $x\sigma = e^{2\pi i x}$

$= \cos 2\pi x + i \sin 2\pi x$, for all $x \in \mathbb{R}$

$x\sigma = 1$, where 1 is the id. element of $\mathbb{C}^*$,

$\Leftrightarrow x \in \mathbb{Z}$,

so that Kernel of σ, $K_\sigma = \mathbb{Z}$.

Also Image of σ is the set of all complex numbers with modules 1, *i.e.*, the set O_1. Thus by Cor. 7.3.1, σ induces an

Mappings in Group Theory 253

isomorphism between O_1 and the factor group $\mathbb{R}|\mathbb{Z}$, i.e., $O_1 \cong \mathbb{R}|\mathbb{Z}$.

Problem 7.7: Let $G = <a>$ be a cyclic group. Prove using fundamental *hmp* theorem, that either $G \cong (\mathbb{Z},+)$ or $G \cong (\mathbb{Z}_m, +_m)$, $m > O$.

Solution: The map $\sigma : \mathbb{Z} \to G$ defined by $n\sigma = a^n$ for all $n \in \mathbb{Z}$ is an epimorphism so that by fundamental *hmp* theorem, $G \cong \mathbb{Z} | K_\sigma$ ...(1), where K_σ is the Kernel of σ

Since $K_\sigma < \mathbb{Z}$ and subgroups of $\mathbb{Z}$ are $\{0\}$, $\mathbb{Z}$ and $H = \{km : k \in \mathbb{Z} \text{ and } m \in \mathbb{Z}^+\}$,

therefore $K_\sigma = \{0\}$ or $\mathbb{Z}$ or H

(i) If $K_\sigma = \{0\}$ (or $\mathbb{Z}$), we have by (1) $G \cong \mathbb{Z}|\{0\} = \mathbb{Z}$, or $G \cong \mathbb{Z}$

(ii) If $K_\sigma = H$, by (1) $G \cong \mathbb{Z}|H$ where $\mathbb{Z}|H = \{0 + H, 1 + H, ..., (m-1) + H\} = \mathbb{Z}_m$, so that $G \cong \mathbb{Z}_m$ $(m > 0)$.

Note: As seen in problem 7.6 and 7.7, using fundamental *hmp* theorem, we can have informations about structures of certain factor groups.

Problem 7.8: If $\sigma : G \to G^1$ is a homomorphism with Kernel K_σ and if $a^1 \in G^1$ such that $a^1 = a\sigma$, $a \in G$, then prove that the set of all those elements of G having image a^1 in G^1 is the left coset aK_σ of K_σ in G.

Solution: Let $a^1 \sigma^{-1}$ be the set of all those elements of G having image a^1, so that

$(a^1) \sigma^{-1} = \{g \in G : g\sigma = a^1\}$

To prove that $(a^1) \sigma^{-1} = a K_\sigma$

Take an arbitrary element x from $(a^1) \sigma^{-1}$,

we get $\qquad x \sigma = a^1$

Now $(a^{-1} x) \sigma$

$\qquad = (a^{-1}) \sigma (x \sigma)$
$\qquad = (a\sigma)^{-1} (x\sigma)$
$\qquad = (a^1)^{-1} a^1 = e^1 \qquad (a\sigma = a^1, \text{given})$

$\Rightarrow$ $\quad a^{-1} x \in K_\sigma \Rightarrow a (a^{-1} x) \in aK_\sigma \Rightarrow x \in aK_\sigma$

Thus $\quad x \in (a^1) \sigma^{-1} \Rightarrow x \in a K_\sigma$

i.e., $\quad (a^1)\sigma^{-1} \subseteq aK_\sigma$ ...(1)

Conversely, take an arbitrary element b from $a K_\sigma$ so that $b = ak$ for some $k \in K_\sigma$ and then $(b) \sigma$

$= (ak) \sigma = (a\sigma)(k\sigma) = (a\sigma) e^1 = a\sigma = a^1$

i.e.,

$(b) \sigma = a^1$ which, by definition, implies that $b \in a^1 \sigma^{-1}$. Thus $b \in a K_\sigma \Rightarrow b \in a^1 \sigma^{-1}$ or $aK_\sigma \subseteq a^1 \sigma^{-1}$ ...(2)

By (1) and (2), $(a^1) \sigma^{-1} = aK_\sigma$.

(Since $K_\sigma \triangleleft G$, $aK_\sigma = K_\sigma a$, and from above result, we have $(a^1) \sigma^{-1} = K_\sigma a$).

Problem 7.9 : If $\sigma : G \to G^1$ is an epimorphism with Kernel K_σ and if H is a subgroup of G then prove that

$$(H\sigma) \sigma^{-1} = H \Leftrightarrow K_\sigma \subset H$$

Solution : (i) Assume that $(H\sigma) \sigma^{-1} = H$

We will prove that $K_\sigma \subset H$

Let $\quad k \in K_\sigma \Rightarrow k \sigma = e^1$ ...(1)

Since $H < G \Rightarrow (H) \sigma < (G) \sigma = G^1$ (here),

$\Rightarrow \quad e^1 \in (H) \sigma \Rightarrow k\sigma \in (H) \sigma$ (by 1)

$\Rightarrow \quad (k\sigma) \sigma^{-1} \in H \Rightarrow k \in H$, or $k_\sigma \subseteq H$.

(ii) Assume that $K_\sigma \subset H$. We will prove that $(H\sigma) \sigma^{-1} = H$

Let $x \in (H\sigma) \sigma^{-1}$ then $x \sigma \in H \sigma$ or $x \in H$ so that $(H\sigma) \sigma^{-1} \subset H$. Conversely, let $x \in H$ then $x \sigma \in H \in \Rightarrow x \in (H\sigma) \sigma^{-1}$, or $H \subset (H\sigma) \sigma^{-1}$. Hence $(H\sigma) \sigma^{-1} = H$.

Problem 7.10 : If a group G is abelian then identity map of G is the only inner automorphism of G and if G is non-abelian then the identity map of G can not be inner automorphism of G.

Solution : (i) Let G be abelian and for a fixed element a of G, σ_a be an inner automorphism. Then for all $x \in G$, $x \sigma_a = a x a^{-1} = a (a^{-1} x) = e x = x$, i.e.,

Mappings in Group Theory 255

$x \sigma_a = x$, for all $x \in G \Rightarrow \sigma_a$ is identity map of G.

(ii) Let G be non-abelian. Then for any $a, b \in G$, $ab \neq ba$
$\Rightarrow (ab)\, a^{-1} \neq (ba)\, a^{-1} \Rightarrow a\, b\, a^{-1} \neq b \Rightarrow b\, \sigma_a \neq b$
or σ_a is not the identity map of G.

Review Exercise on Chapter 7

1. Find Kernel of following homomorphism :
   (i) $\sigma_1 : (\mathbb{Z}, +) \to (Q^*, \cdot)$ given by $m\sigma = 7^m$ for all $m \in \mathbb{Z}$.
   (ii) $\sigma_2 : (\mathbb{Z}, +) \to G, \cdot)$ given by $m\sigma = x^m$, for all $m \in \mathbb{Z}$.
   Here $(G, \cdot)$ is any group and $x \in G$.

2. Prove that the map $g : S_n \to (\{\pm 1\}, \cdot)$ defined by
$$ag = 1 \quad \text{if } a \in A_n$$
$$ag = -1 \quad \text{if } a \in S_n - A_n$$
is an epimorphism. Find K_g and check the fundamental *hmp* theorem that $\{\pm 1\} \cong S_n | K_g$.

3. Define an onto *hmp* $f : (\mathbb{Z}, +) \to (\mathbb{Z}_5, +_5)$ and find its Kernel K_f. Verify the fundamental *hmp* theorem that
$$\mathbb{Z}_5 \cong \mathbb{Z} | K_f$$

4. If the homomorphism $\sigma : G \to G^1$ is an isomorphism as well. prove that there exists an *hmp* $\pi : G^1 \to G$ such that $\pi \, 0 \, \sigma = \pi \sigma = I_G$ and $\sigma \, 0 \, \pi = \sigma \pi = I_G$.

5. Let G be a group and $H < G$. Then prove that $H \triangleleft G$ if and only if H is Kernel of some homomorphism.

(*Hint* : Follow Theorem 7.3.6)

6. Let $(G, \cdot)$ be group of all $n \times n$ invertible matrices with elements of G from $\mathbb{R}$ and $G^1 = (\mathbb{R}^*, \cdot)$. Then prove that the function $f : G \to G^1$ defined by $Af = |A|$, for all $A \in G$, where $|A|$ denotes the determinant of A is an epimorphism but not an isomorphism. Prove also that the Kernel K_f = Set of all such matrices with determinant equal to 1.

7. Let $\sigma : G \to G^1$ be an onto homomorphism, then prove the following :

(i) $H^1 < G^1 \Rightarrow H = (H^1) \sigma^{-1} < G$ and $K_\sigma \subseteq H$

(ii) $H^1 \triangleleft G^1 \Rightarrow H = (H^1) \sigma^{-1} \triangleleft G$

(iii) If $K \supseteq K_\sigma$ where $K < G$ such that $(K) \sigma = H^1$ then $K = H$

8. Let $\sigma : G \to G^1$ be a *hmp*, $N \triangleleft G$ and $N^1 \triangleleft G^1$ such that $(N) \sigma \subseteq N^1$. Then prove that there exists a *hmp* $f : G|N \to G^1|N^1$ Deduce that if $K_\sigma = N$ then there exists a monomorphism $g : G|N \to G^1$.

(*Hint* : Define $if : G|N \to G^1|N^1$ by the rule $(aN)f = (a\sigma) N^1$ for all $a \in G$ and $a N \in G|N$.)

9. Let G be a group and σ be an automorphism of G, then prove that $N \triangleleft G \Rightarrow (N) \sigma \triangleleft G$.

10. Let $G = <2\pi>$ be a cyclic group under addition. Prove using fundamental *hmp* theorem that

$(\mathbb{R}, +)|G \cong (O_1, \cdot)$, where $(O_1, \cdot)$ is the group of all complex numbers of modules 1 under multiplication.

APPENDIX

Exact Sequences of Groups

In this appendix, we will briefly discuss exact sequences of groups, a concept which is generated by considering homomorphisms between three or more groups. The objective is to provide the readers an insight into the general situation of series of Groups with which they will come across in more higher topics of Groups.

Definitions

(1) The sequences of groups G_1, G_2, G_3 alongwith homomorphism $\alpha : G_1 \to G_2$ and $\beta : G_2 ; G_3$, written as

$G_1 \xrightarrow{\alpha} G_2 \xrightarrow{\beta} G_3$, is called an exact sequence at the middle group G_2 iff Image of α = Kernel of β, or $(G_1)\alpha = K_\beta$

Mappings in Group Theory

(2) In general, a finite or infinite sequence of groups

$$\ldots \to G_{i-1} \xrightarrow{\alpha_i} G_i \xrightarrow{\alpha_{i+1}} G_{i+1} \xrightarrow{\alpha_{i+2}} G_{i+2} \to \ldots$$

alongwith the *hmps* $\alpha_i, \alpha_{i+1}, \alpha_{i+2}, \ldots$ is said to be an exact sequence if each of its subsequences of three consecutive terms is exact at the middle group.

Thus above sequence of groups is exact iff (i) Image of $\alpha_i = K_{\alpha_{i+1}}$, (ii) Image of $\alpha_{i+1} = K_{\alpha_{i+2}}, \ldots$ etc.

(3) An one element group is called a *trivial group* and is denoted by O. Thus $O = \{e\}$.

(4) Any exact sequence of groups in which the first and the last terms are the trivial groups is called a *short exact sequence*. Thus the exact sequence

$$O \to G_1 \to G_2 \to G_3 \to O$$

is a five term short exact sequence.

(5) A homomorphism $\alpha : G \to G^1$ is called a *trivial hmp* if α is a constant map.

Thus the *hmp* $\alpha : G \to G^1$ is trivial if

(i) Image of α, (G) $\alpha = \{e^1\}$, and

(ii) Kernel of α, $K_\alpha = G$, the entire domain.

Examples

1. Consider the following sequence of groups

$$(2\mathbb{Z}, +) \xrightarrow{\alpha} (\mathbb{Z}, +) \xrightarrow{\beta} (Q^*, \cdot)$$

where the *hmps* α and β are defined by

$$m\alpha = m. \text{ for all } m \in 2\mathbb{Z}$$

and $n\beta \qquad = (-1)^n, \text{ for all } n \in \mathbb{Z}$

Then,

Image of α, $(2\mathbb{Z}) \alpha = 2\mathbb{Z}$, and

Kernel of β, K_β = Set of all those elements of $\mathbb{Z}$ which, by β, are mapped to 1, the id. element of $(Q^*, \cdot)$

$= \{\ldots -4, -2, 0, 2, 4, \ldots\}$
$= 2\mathbb{Z}$

Thus $(2\mathbb{Z})\alpha = K_\beta$, i.e.,

Image of α = Kernel of β, so that the given sequence is an exact sequence.

(2) Consider the sequence

$N \xrightarrow{I} G \xrightarrow{\nu} G|N$, where $N \triangleleft G$, G is any group, I is identity map on N and ν is natural *hmp*. Here $nI = n$ for all $n \in N$ so that

Image of I, (N) I = N. By definition of natural (canonical) *hmp* ν, $K_\nu = N$

Thus

(N) $1 = K_\nu$, i.e., Image of I = Kernel of ν or the sequence is exact.

Questions

1. (1) If the sequence of groups $(3\mathbb{Z}, +) \xrightarrow{\alpha} (\mathbb{Z}, +) \xrightarrow{\beta} (2\mathbb{Z}, +)$ is exact, and if $n\beta = 2n$ for all $n \in \mathbb{Z}$, find the map α.

Solution : Since sequence is exact, Image of α = Kernel of β

$\Rightarrow$ $(3\mathbb{Z})\, \alpha = \{0\}$, as $K_\beta = \{0\}$

or $(3\mathbb{Z})\, \alpha = \{e^1\}$, where e^1 is the id-element of $(\mathbb{Z}, +)$

$\Rightarrow$ The map α is a trivial *hmp*.

(2) In the following exact sequence of groups $G_1 \xrightarrow{\alpha} G_2 \xrightarrow{\beta} G_3$, find the image of element $g_1 \in G_1$ under the composite map $\alpha\beta$.

Solution : Since sequence is exact,

Image of $\alpha = K_\beta$

$\Rightarrow (G_1)\, \alpha = K_\beta$

Thus $g_1 \in G_1 \Rightarrow g_1 \alpha \in K_\beta \Rightarrow (g_1 \alpha) \beta = e_3$, where e_3 is id. element of G_3, i.e., $(g_1) \alpha\beta = e_3$ or image of $g_1 \in G_1$ under map $\alpha\beta$ is the element e_3.

(3) If the short sequence $O \xrightarrow{\alpha} G \xrightarrow{\beta} O$ is exact, find the group G.

Solution : Since sequence is exact,

Image α = Kernel of β

$\Rightarrow$ (O) α = K_β

$\Rightarrow$ $\{e\}$ = G (K_β is the entire set G)

Thus G is one element group.

Theorems : (I) If the short term sequence

$O \xrightarrow{\alpha} G_1 \xrightarrow{\sigma} G_2 \xrightarrow{\beta} O$ is exact then $G_1 \cong G_2$.

Proof : The sub sequences $O \xrightarrow{\alpha} G_1 \xrightarrow{\sigma} G_2$ and $G_1 \xrightarrow{\sigma} G_2 \xrightarrow{\beta} O$ are exact at their respective middle groups G_1 and G_2.

Thus

$\qquad$ Image of $\alpha = K_\sigma$ $\qquad$ ...(1), and

$\qquad$ Image of $\sigma = K_\beta$ $\qquad$ ...(2)

By (1), $\{e_1\}$ = K_σ where e_1 is id. element of G_1

$\Rightarrow$ σ is monomorphism $\qquad$ (by Th. 7.3.3)

By (2), Image of σ, (G_1) $\sigma = G_2$,

i.e., $\sigma : G_1 \to G_2$ is onto and hence an epimorphism.

Thus $G_1 \cong G_2$

(II) If the short term sequence

$O \xrightarrow{\alpha} A \xrightarrow{\alpha} B \xrightarrow{\beta} C \xrightarrow{\mu} O$

is exact then

(*i*) the map $\alpha : A \to B$ is monomorphism

(*ii*) the map $\beta : B \to C$ is epimorphism

$\qquad$ and

(*iii*) $C \cong B\,|\,(A)\,\alpha$

Proof : Since the given five term short sequence is exact, each of the following sub sequences of three consecutive terms are exact at their respective middle groups :

$O \xrightarrow{\sigma} A \xrightarrow{\alpha} B$, $A \xrightarrow{\alpha} B \xrightarrow{\beta} C$ and $B \xrightarrow{\beta} C \xrightarrow{\mu} O$.

Thus

$$\text{Image of } \sigma = K_\alpha \qquad \ldots(1)$$
$$\text{Image of } \alpha = K_\beta \qquad \ldots(2)$$
$$\text{and Image of } \beta = K\mu \qquad \ldots(3)$$

By (1), $\{e_A\} = K_\alpha$ or the map $\alpha : A \to B$ is monomorphism.

By (2), $\qquad (A)\alpha = K_\beta \qquad \ldots(4)$

By (3), $(B) b = C$ as $K\mu$ is the entire group C

$\Rightarrow$ the map $\beta : B \to C$ is an epimorphism

Applying fundamental *hmp* theorem in the epimorphism $\beta : B \to C$, we get

$$C \cong B | K_\beta$$
$\Rightarrow \qquad C \cong B | (A)\alpha \quad \text{(by 4)}$

(III) In the following four term exact sequence

$G_1 \xrightarrow{\alpha_1} G_2 \xrightarrow{\alpha_2} G_3 \xrightarrow{\alpha_3} G_4$, the following statements are equivalent :

(i) α_1 is an epimorphism

(ii) α_2 is a trivial hmp

(iii) α_3 is a monomorphism

Proof : We will prove that

$$(i) \Rightarrow (ii) \Rightarrow (iii) \Rightarrow (i)$$

to show that the above statements are equivalent. By exactness of the subsequences $G_1 \xrightarrow{\alpha_1} G_2 \xrightarrow{\alpha_2} G_3$ and $G_2 \xrightarrow{\alpha_2} G_3 \xrightarrow{\alpha_3} G_4$, we have

$$\text{Image of } \alpha_1 = K_{\alpha_2} \qquad \ldots(1), \text{ and}$$
$$\text{Image of } \alpha_2 = K_{\alpha_3} \qquad \ldots(2)$$

$(i) \Rightarrow (ii)$:

By (1), $(G_1)\alpha_1 = K_{\alpha_2}$ But by given statement (i) that α_1 is an epimorphism, $(G_1)\alpha_1 = G_2$

$\Rightarrow K_{\alpha_2} = G_2 \Rightarrow$ the map $\alpha_2 : G_2 \to G_3$ is a trivial homomorphism.

(ii) $\Rightarrow$ (iii) : We are given that
$\alpha_2 : G_2 \to G_3$ is a trivial hmp
$\Rightarrow$ Image of α_2, $(G_2)\alpha_2 = \{e_3\}$,

where e_3 is id. element of group G_3. Thus by (2), $K_{\alpha_3} = \{e_3\}$ so that the map $\alpha_3 : G_3 \to G_4$ is a monomorphism.

(iii) $\Rightarrow$ (i) : We are given that the map $\alpha_3 : G_3 \to G_4$ is monomorphism

$\Rightarrow K_{\alpha_3} = \{e_3\}$

By (2), $(G_2)\alpha_2 = K_{\alpha_3} \Rightarrow (G_2)\alpha_2 = \{e_3\}$

$\Rightarrow \alpha_2 : G_2 \to G_3$ is trivial $\Rightarrow K_{\alpha_2} = G_2$, which by (1) implies that image of $\alpha_1 = G_2$

$\Rightarrow (G_1)\alpha_1 = G_2$ so that the map
$\alpha_1 : G_1 \to G_2$ is an epimorphism.

Exercise

(1) In the following five term exact sequence

$G_1 \xrightarrow{\alpha_1} G_2 \xrightarrow{\alpha_2} G_3 \xrightarrow{\alpha_3} G_4 \xrightarrow{\alpha_4} G_5$, if the group G_3 is trivial ($G_3 = O$) then prove that : (i) α_1 is an epimorphism, and (ii) α_4 is a monomorphism.

(2) In the following exact sequence of groups

$G_1 \xrightarrow{\alpha_1} G_2 \xrightarrow{\alpha_2} G_3 \xrightarrow{\alpha_3} G_4 \xrightarrow{\alpha_4} G_5 \xrightarrow{\alpha_5} G_6$

show that the following statements are equivalent :

(i) α_3 is an isomorphism

(ii) α_2 and α_4 are trivial homomorphism

(iii) α_1 is an epimorphism and α_5 is a monomorphism.

BIBLIOGRAPHY

1. Childs, L.N., A Concrete Introduction to Higher Algebra, Springer Verlag, Second edition (1995).
2. Cohen, A.M., Cuypers, H. and Sterk, H., Algebra Interactive!, Springer Verlag (1999).
3. Fraleigh, J.B., A first course in Abstract Algebra, Narosa Publishing House, New Delhi (1992).
4. Herstein, I.N., Topics in Algebra, Vikas Publishing House Pvt. Ltd., New Delhi (1985).
5. Hungerford, T.W., Algebra, Rinehart & Winstan Inc., New York (1976).
6. Jacobson, N., Basic Algebra, Vol. I and II, Freeman, San Francisco, (1980).
7. Mac Donald, D., The Theory of Groups, Oxford (1968).
8. Rodes, F., On the Fundamental Group of a Transformation Group, Proc. London Math. Soc., 16 (1966) 635-50.
9. Rotman, J.J., An Introduction to the Theory of Groups, Iowa, Wm-C. Brown (1988).
10. Sagan, B.E., The Symmetric Group, Second edition, Springer-Verlag (2000).